HEALTH PHYSICS
OPERATIONAL MONITORING

HEALTH PHYSICS
OPERATIONAL MONITORING

Editors

CHARLES A. WILLIS
Potomac Electric Power Co.

JOHN S. HANDLOSER
E. G. & G., Inc.

VOLUME 2

GORDON AND BREACH, SCIENCE PUBLISHERS
NEW YORK • LONDON • PARIS

Library of Congress Catalog Card Number: 78-188888

Editorial office for Great Britain
    Gordon and Breach, Science Publishers Ltd.
    41-42 William IV Street
    London, W.C. 2, England

Editorial office for France
    Gordon and Breach
    7-9 rue Emile Dubois
    Paris 14^e, France

ISBN 0 677 13670 6 (cloth); ISBN 0 677 13675 7 (paper)

Printed in the United States of America

PREFACE

Operational monitoring is the essential part of health physics. While research and the other more esoteric aspects are important, the mainstay of the health physics profession is operational monitoring. The importance of the subject matter combined with the lack of any other comprehensive reference on the subject provides a special impetus for the publication of these transactions.

The principal purposes of this publication are, first, to disseminate applied health physics information; second, to provide a basic reference document for field health physicists; third, to provide a convenient way of introducing the health physics student to operational problems, and finally, to promote the writing of a textbook on applied health physics. While the shortcomings of a symposium-proceedings book are obvious, it is hoped that these volumes will meet the outstanding immediate needs, at least until the long-sought applied health physics text appears.

Partially as a result of the volume of material to be accommodated, the discussions associated with each paper have been omitted. In several ways this is regrettable. Questions and answers are often both interesting and informative. Further, comments might alert the reader and remind him that the various papers represent the views of the authors and not necessarily the position of the health physics profession. For example, Dr. Wegst's comments on the public relations constraints on communications met with objections from Mr. R. L. Kathren. Similarly, Dr. K. Z. Morgan voiced some noteworthy objections to Mr. C. J. Sternhagen's evaluation of uranium mining hazards. There was, in fact, considerable spirted discussion which cannot be included here. It is hoped that these comments will be made available through the journals.

This book and the meeting it reports were made possible by the support of many people. The assistance and encouragement of the Health Physics Society directors, particularly President Langham, were vital. The unflagging efforts of the members and directors of the Southern California Chapter and the committee members in particular, are gratefully acknowledge. This publication, from the selection of the papers through compilation and correction of the proofs, has largely been the responsibility of Mr. John Handloser and the Program Committee. Ultimately, of course, any success enjoyed by this book is due to the contributors.

Charles A. Willis, General Chairman

Session III - FACILITY MONITORING

Session IV - EXPOSURE EVALUATION

Session VIII - EDUCATION AND TRAINING

Session IX -EMERGENCY PLANNING AND EXPERIENCES

Session V
INSTRUMENTATION

Chairmen

WILLIAM SAYER
Atomics International
Canoga Park, Calif.

DALE E. HANKINS
Los Alamos Scientific Laboratory
Los Alamos, New Mexico

INSTRUMENTS IN THE FIELD
USE, ABUSE, AND MISUSE*

R. L. Kathren
Battelle Memorial Institute
Pacific Northwest Laboratory
Richland, Washington

ABSTRACT

Advances in field instrumentation have been great
over the past decade, providing the operational health
physicist with a powerful armamentarium. However,
even the most modern instruments have many limitations, and measurements must be made and evaluated
in the light of these limitations, which include energy
and angular dependence, response to pulsed sources,
and effects of mixed radiation fields. The concept of
surface dose rate, its meaning, use, and misuses
will be examined in the context of operational health

*This paper is based on work performed under
United States Atomic Energy Commission Contract
AT(45-1)-1830;

physics. A practical method of measuring doses close
to the source—air interface will be discussed.

I. THE EVOLUTION OF PORTABLE
SURVEY INSTRUMENTS

Few professions are as dependent on field survey
instruments as is health physics. While much can be
done without instruments, operational health physics
would be virtually non-existent without portable survey meters, for the operational health physicist needs
to have meaningful measurements of conditions in the
field in order to intelligently gauge the success of his
efforts.

Although radiation protection efforts began
shortly after the discovery of x-rays and radioactivity,
development of field monitoring instrumentation lagged
far behind other efforts. Photographic methods were
the first to be applied to field measurements of ionizing radiation. Perhaps the very first attempts at
field monitoring for protection purposes were those
of William Rollins, an early ancestor of today's
health physicist. In 1902, William Rollins described
a method of shielding x-ray tube enclosures and prescribed the use of photographic plates placed against
the housing to monitor leakage.[1] If the plate was not
fogged by an exposure time of seven minutes, the
tube housing was presumably satisfactory. A similar
application for photographic film was described a year
later by an American dermatologist who was concerned
with limiting exposure to his patients.[2] In this case,
the film was placed on the patient to directly monitor
the exposure.

Film badges for personnel monitoring evolved during the 1920's,[3,4] but portable survey instruments were still lacking. What was quite possibly the first true portable survey meter was designed and built in 1929 by L. S. Taylor.[5] This interesting device had three interchangeable spherical aluminum ionization chambers, each of a different size to provide several sensitivity ranges. Three small B batteries provided the 135 volt potential for the chamber. Charge was measured by a string electormeter, and full scale with the largest (most sensitive) chamber was 2 mR. Most significantly, this instrument was designed exclusively for protection purposes.

Through the 1930's, although x-rays and radium were in wide use, scant attention was paid to development and manufacture of portable radiation survey meters. However, at least one commercial manufacturer produced a portable ionization chamber, and the invention of the Geiger-Mueller tube in 1928 lead to the first commercial Geiger counters, available in the middle and late '30's.

However, it was not until the Manhattan District that any real attention was paid to the design, development, and application of portable monitoring instruments. The Health Division assumed principal responsibility for the development of pocket dosimeters and film badges, and cooperated with the Physics Division in creating portable survey meters, with one of the more notable efforts being "Pluto," an alpha contamination survey meter.[6] Operational health physics, born along with the atomic bomb, brought with it a need for portable survey instruments, and the hiatus between protection efforts and portable monitoring instruments and development had begun to close.

The late '40's saw the birth of many new portable instruments. Big, bulky, and heavy by modern standards, these nonetheless were vital to those working in health physics. Serious development efforts were begun, with emphasis on miniaturization and improved response. The transistor, discovered in 1947, was swiftly incorporated into radiation monitoring instruments, and along with the more efficient batteries developed during the early 1950's, permitted great reduction in size and weight. A variety of new detectors appeared—thin metal walled Geiger tubes, pressurized ionization chambers, and neutron detectors of various types all appeared in commercially available instruments.

The flavor of portable survey meter evolution can be captured by a brief look at the history of a single instrument species. The Geiger counter, notable as one of the lay symbols of the nuclear era might be a fitting choice. In the late 1940's, several Geiger counters were available commercially. Large and bulky, these featured glass wall tubes and were highly energy dependent ~ so much so that they were often considered detectors rather than dose rate meters. The vacuum tube circuitry required relatively high current, and big, heavy batteries were the order of the day. Saturation—or downscale reading in high radiation fields—was a severe bugaboo. These forerunners of modern day GM survey meters typically weighed in at 10-15 lbs., and displaced several hundred cubic inches. Typical of these early GM's were the Nuclear Chicago Model 2610, the Victoreen Model 263, and the first of the military AN/PDR 27 series;

those who have used any of these instruments cannot fail to recall their idiosyncrasies, weight, and size.

The early 1950's brought the application of the transistor to the Geiger counter, and the first metal wall GM tubes made their appearance. Energy dependence and saturation were still problems, but bulk and weight had been cut in half. In these transitional years, vacuum tubes were still used, as were heavy B batteries, and instruments still weighed around ten pounds. Glass wall tubes were still prevalent, and many a high reading was attributable to a light leak. But the pattern for lightweight, small packages was beginning to emerge. Representative instruments for the early '50's include the Precision Model 107, and the early Civil Defense type units.

In the mid-1950's, the changes introduced in the early years of the decade became more prevalent. Metal wall, halogen quenched tubes were common and energy dependence was dramatically improved, partly by the use of low Z metal wall tubes, and partially by improved detector housings. Transistors made possible a very small package, and the lower current drain permitted the use of two D cells as the entire battery supply. Volume and weight were reduced even further—the typical GM survey meter of a decade ago weighed only a few pounds, and is exemplified by the Eberline E-112B, and the military AN/PDR 27J, a fully transistored light weight successor to the earlier models of the '27 series.

In the early years of this decade, the saturation problem appeared to have been licked, and an even more exciting development was current mode operation of a GM tube, which extended the range by two to three orders of magnitude. The detection versus dose rate controversy was, for all practical purposes, solved—most manufacturers labeling the meter scale in units of mR/hr. Miniaturization reached a practical limit and beyond, for now, GM survey meters seem to be getting larger.

II. USE, ABUSE, AND MISUSE

Survey meters have evolved over the years into highly sophisticated and useful devices. Transistors, integrated circuits, MOS-FETs—these, and other improvements have given the operational health physicist of 1969 tools beyond the ken of his forerunners of only a generation ago. Yet in spite of advances in instrument technology, many of the old problems are still with us, compounded by new nuclides and radiation generating machines providing unique energy and intensity spectra.

Use of field survey instruments is an easy task, since virtually anyone can turn an instrument on and read the meter response. But proper use is another story, and implies judgement as well as knowledge of instrument response characteristics.

Areas of misuse are many, and no short presentation could attempt to enumerate them. However, most instances of instrument abuse and misuse can be compartmentalized into four general areas:

    a. Energy dependence

    b. Mixed field effects

    c. Pulsed field response

    d. Angular dependence and other geometry considerations

Each of these will be briefly considered in turn.

With the exception of a few tissue equivalent chambers, virtually all instruments are energy dependent. The problem is neither new or is it limited to photon monitoring. However, most of the abuse occurs in the monitoring of low energy photons. Two anecdotes will illustrate.

About a decade ago, while a very junior health physicist, I witnessed the use of a high range scintillator for monitoring a laboratory x-ray machine. The scintillator, held in the primary beam by a senior health physicist, hardly responded to the maximum output of the 50 kVp machine. Small wonder, for the detector was located behind a quarter inch of steel. On the basis of this measurement, the health physicist concluded that a frank x-ray burn on the hand of a chemist could not have occurred! The second event took place about a year ago, when I was astonished to see in a popular publication an evaluation of leakage radiation from color television sets.[7] That there was leakage of x-rays was not surprising, but the mode of measurement was. Direct meter readings from thin end window GM survey meters were used to quantify x-rays in the region below 25 keV.

These two anecdotes serve to illustrate flagrant misuse of an instrument--using it outside of its design capabilities, with no attempt at interpreting or correcting the readings. Moreover, the irresponsible practice of health physics in one case, and the misleading of the public in the other, is intolerable.

The problem of improper evaluation is not limited to low energy photons. Survey meters readings are routinely used without interpretation for photons above 5 MeV, in flagrant disregard of electronic equilibrium

conditions. Neutron survey meters, the accuracy of
which is notably poor, are used with total and utter
disregard of the neutron spectrum and detector response. Even the new "rem meters" are not energy
independent.

Mixed field effects constitute a relatively minor
area of misuse. However, as newer and larger accelerators are built, mixed field monitoring will become increasingly important. Although in most instances, the error caused by a mixed field will be one
of over-estimation, it is nonetheless an error, and
could lead to unnecessary expense if used as a basis
for, say, shielding design. Such use, while admitted-
ly poor practice, is not unrealistic, for shielding design must ordinarily be verified by actual field measurement.

Pulsed fields are rapidly assuming greater significance. Pulsed reactors, accelerators, and x-ray
generators are commonplace, and most field survey
instruments cannot be used for monitoring the high
intensity, short duration pulses associated with these
devices. Even the venerable film dosimeter has its
upper intensity limits.[8]

Most instruments will correctly respond to fields
with frequencies greater than a few pulses per second.
Many x-ray machines and cyclotrons are, in fact,
pulsed, but the frequency is so great that the instrument sees essentially a constant field. At lower pulse
frequencies, on the order of one or two pulses per
second, most survey meters become virtually useless. However, ion chambers can be useful if the time
constant is sufficiently long, for after a period of
buildup, the reading will level out, and reflect averaging of the dose rate. A Juno, for example, was

successfully used for surveying around a pulsed accelerator with repetition rates as low as 30 per minute. The validity of the Juno readings was verified with LiF thermoluminescence dosimeters.

The final area of abuse concerns the geometry of the chamber and source. Angular dependence of detectors is all but ignored, an abuse that could seriously underestimate the dose rate in an ambient field. Other geometry considerations relating to source and detector size will be considered for a representative instrument in a later paper in this session.[9] Suffice to say that the total solid angle intercepted by the chamber is the one of the basic criteria, and suitable corrections must be made to avoid underestimates of dose rate. But by far, the most prevalent and pernicious abuse of instrument capability relates to attempts to monitor surface dose rate.

III. WHAT IS SURFACE DOSE RATE ?

Surface dose rate is, in simple terms, the dose rate at the interface of a surface emitting radiation and the air. Conceptually, it is the maximum dose rate that an individual could incur were he to handle a source--bare or packaged. Calculation and measurement of surface dose rate is difficult, at best; surface dose rates cannot be determined by the simple expedient of laying the detector on the surface, or in contact with the source. Yet to many, this is precisely how to obtain the surface dose rate. In this context, the terms "surface dose rate' or 'contact dose rate' are meaningless; obviously, different instruments will give different fractions of the true

value. Less obvious, but of great importance, is
that distance alone is not the only factor affecting the
response of the instrument. Even in the case of a
large plane source, much greater in cross-sectional
area than the detector, other factors will affect the
measurement, including chamber wall material,
average electrical field strength within the detector,
shape of the detector, and source geometry and variations in uniformity.

Rather than give a 'contact' or 'surface' reading,
a more pragmatic approach might be to determine the
relationship of surface dose rate to the dose rate at
some convenient distance from the surface. Then,
the instrument could be used to determine the dose
rate at the convenient distance and correction factor
applied to give an estimate of surface dose rate.
Preliminary experiments conducted in our laboratory
with relatively large plane beta sources (diameter >
4") indicate that the ratio of surface dose rate to dose
rate at 10 cm from the surface is constant and approximately equal to 25. Beta doses were determined with
an extrapolation chamber, which enables the surface
dose rate to be measured. Although, as yet, the data
are too preliminary for general application, the concept appears to be promising. A similar ratio has
been noted with large plane photon emitting sources.

IV. A RAY OF HOPE

As health physicists, we have a professional
obligation to properly use our instruments. However,
available information regarding instrument response
is incomplete, and what is available is often inaccurate

or only partially correct, encouraging misuse and error. Manufacturers of health physics instrumentation could help if they would establish and adhere to standard means of measuring and reporting instrument response. A comprehensive instrument evaluation program will be described later in this session;[10] perhaps some of the ideas put forth will merit implementation.

In many respects, the problems of nuclear instrument specification statements is analagous to that faced by the high fidelity instrument manufacturers a decade ago. At that time, each manufacturer used his own method of measuring and reporting the response of his equipment. Hence, one manufacturer's 25 watts was another's 100 watts, and each was correct in that he was reporting the results of his tests. However, power output can be measured in many different ways, and, over the years, the Institute of High Fidelity (IHF) has issued various standards prescribing measurement and testing techniques, as well as definitions. While not ideal, the current IHF Standards for audio amplifier tests and measurements (IHF-A-201 [1966]) and tuner measurements (IHFM-T-100 [1958]) do provide a common reference point. Manufacturers using these IHF Standards are required to present a specified minimum of published specifications, all obtained according to the IHF Standard. In this manner, the user has some knowledge of the performance of his instruments as determined by a series of standardized testing procedures.

Were a similar scheme established by the manufacturers of health physic instruments, misuse and abuse of instrument capabilities would be curtailed. While complete instrument evaluation is desirable, it may not be practicable. But, perhaps some common ground and self-regulation are indicated. The implications of an erroneously used and interpreted survey meter are indeed serious.

REFERENCES

1. W. Rollins, "Non-Radiable Cases for X-Light Tubes," Elect. Rev. 40:795 (1902).

2. S. Stem, "Method for Measuring the Quantity of X-Ray," J. Curtaneous Dis. 21:568 (1903).

3. E. H. Quimby, "A Method for the Study of Scattered and Secondary Radiation in X-Ray and Radium Laboratories," Radiology 7:211 (1926).

4. R. S. Landauer, "The Use of Dental Films in the Determination of Stray Radiation," Radiology 8:512 (1927).

5. L. S. Taylor, "An Early Portable Radiation Survey Meter," Health Physics 13:1347 (1967).

6. H. D. Smyth, Atomic Energy for Military Purposes, Princeton University Press, Princeton, (1945), pp. 150-152.

7. "Large-Screen Color TV," Consumer Reports, (January, 1968), pp. 16-20.

8. E. Tochilin and N. Goldstein, "Dose Rate and Spectral Measurements from Pulsed X-Ray Generators," Health Physics 12:1705 (1966).

9. W. P. Howell and R. L. Kathren, "Calibration and Field Use of Ionization Chamber Survey Instruments," BNWL-SA-2096; Presented at the Health Physics Society Midyear Symposium on Operational Monitoring (January, 1969).

10. L. V. Zuerner and R. L. Kathren, "Evaluation Program for Portable Radiation Survey Meters," BNWL-SA-1947; Presented at the Health Physics Society Midyear Symposium on Operational Monitoring (January, 1969).

EVALUATION PROGRAM FOR PORTABLE RADIATION MONITORING INSTRUMENTS

L. V. Zuerner, R. L. Kathren
Battelle Memorial Institute
Pacific Northwest Laboratory
Richland, Washington

Knowledge of response and performance capabilities is of practical importance in selection and use of radiological measuring devices for health physics purposes. Instruments are desired which will meet prescribed performance specifications with the lowest overall cost per unit of time. In some cases, a high initial cost will be more than offset by lower operating and maintenance charges.

A well oriented and directed evaluation program should examine appropriate physical, electronic, and radiological characteristics of the instrument, providing data for performance and cost analysis. For

Work performed under Contract Number AT(45-1)-1830 between the U. S. Atomic Energy Commission and Battelle Memorial Institute.

portable instruments, weight, strength, and ease of
handling and servicing are important physical characteristics. Human engineering features such as ease
of meter reading, availability of controls, and physical stability must also be considered.

Among the electronic features to be considered
are sensitivity, temperature and voltage dependence,
stability, battery lifetime, response to electrical
interference, noise and vibration, and extracameral
radiation. Radiological factors include accuracy and
stability of calibration, energy dependence, response
to unwanted radiations, temperature dependence, and
saturation effects.

A typical evaluation program is outlined for portable radiation survey meters and recommendations
made for a standardized system of determining and
reporting instrument performance.

1.0. INTRODUCTION

Proper selection of portable radiation monitoring
instruments is as important as proper use. The ultimate objective, of course, is to procure instruments
that will meet prescribed performance specifications
with the lowest overall cost per unit.

Selection implies prior evaluation, and a meaningful evaluation must be based on knowledge of the
performance characteristics, limitations, and economics of the instrument under consideration. Unfortunately, adequate and accurate data on which to
base a meaningful evaluation is unavailable in most
instances. Perhaps the best available are the published
specifications of the manufacturer, but these are often

incomplete or couched in language that is not wholly comprehensible, and may even be misleading. And, since each manufacturer may report or measure the same performance characteristics differently, the published specifications of various manufacturers may not be comparable. Many instrument purchases, however, are based solely upon the claims of the manufacturer or his representative, published or otherwise.

A second body of relatively accessible data is word-of-mouth from those who have had prior experience with a given instrument. In this case, the potential for misinformation is quite high; the spoken word gets notoriously garbled in transmission through several persons. And, even a direct two party contact may not always provide accurate information, either by accident or intent; some, we all realize, will not admit to purchasing a poor instrument.

What is needed is a well oriented and direct evaluation program to determine appropriate physical, electronic, and radiological characteristics on which a performance and cost evaluation can be based. A typical evaluation program for portable radiation survey meters is outlined in Table I. Obviously, no single program would fit all portable instruments now in use, or even those commercially available. However, the skeletal program in Table I should provide a sufficient body of information to enable the health physicist to make a sound selection, consistent with the needs of his operation, and the dictates of his budget.

As shown in the Table, the evaluation of a portable survey meter can be considered in the context of three broad areas: mechanical, electronic, and

Table I

Typical Evaluation Areas for Portable
Radiation Survey Meters

I. MECHANICAL

  A. Physical Construction
  B. Shock and Moisture Resistance
  C. Human Factors Engineering

II. ELECTRONIC

  A. Power Supply
      1. Stability
      2. Temperature Dependence
      3. Battery Life
  B. Input Sensitivities
  C. Linearity
  D. Electromagnetic Interference
      1. Magnetic Fields
      2. AC Induced Fields and Transients
      3. Radio Frequency
      4. Electrostatic Fields
  E. Switching Transients
  F. Capacitance Effects
  G. Geotropic Effects
  H. Temperature Dependence
  I. Extracameral Effects
  J. Sound and Vibration Effects

III. RADIOLOGICAL

  A. Range
  B. Sensitivity and Detection Limit
  C. Accuracy

Table I (Cont'd.)

D. Reproducibility
E. Saturation
F. Energy Dependence
G. Temperature and Pressure Dependence
H. Angular Dependence
I. Response to Unwanted Radiations

radiological. Each of these areas will be discussed in turn, and a series of specific evaluation and measurement procedures presented.

2.0. MECHANICAL EVALUATION

Evaluation of the mechanical features of an instrument can be divided into two overlapping areas: physical construction and human engineering. To a relatively large degree, this portion of the evaluation is subjective, and requires the sound judgment that is often acquired only through experience.

2.1. PHYSICAL CONSTRUCTION

Physical construction is evaluated by general appearance of both exterior and interior, paying particular attention to quality control and workmanship. Solder joints are of particular value in this evaluation; the presence of cold joints, excessive amounts of solder or flux, or the lack of good mechanical coupling are indicative of poor quality. Circuit boards are similarly useful indicators; the quantity of path conductor and its path length, sharpness of the etching, and labeling of components are points to be considered. The examination of component layout should include consideration of heat dissipation and length of leads and connectors.

2.2. SHOCK AND MOISTURE RESISTANCE

While evaluation of structural strength is primarily subjective, resistance to shock and moisture can be objectively examined. Field instruments generally should have the ability to withstand a drop of three feet onto a hard surface and should function properly regardless of ambient humidity. Moreover, they should be able to survive a single, rapid total immersion in water without ill effects. These simple checks can reveal a great deal about the construction of an instrument; the detector, of course, should generally be exempted from meeting these requirements.

2.3. HUMAN ENGINEERING

An important aspect of portable survey instruments, often overlooked, is human engineering. Again, the evaluation is essentially subjective, and should include consideration of safety hazards, along with ease of handling, readout, and servicing. Typical safety hazards include sharp edges, inadequate grounding, and other shock hazards.

The shape, size, and weight of the instrument are important and obvious factors in ease of handling, and are seldom overlooked. However, other more subtle human engineering features also need to be included in the total evaluation--the location of switches, switching arrangements, meter size, location and readability—these are all points to consider. Even the general external appearance of an instrument can affect its acceptance by monitoring and other field personnel. Some consideration should also be given to use of portable instruments by personnel wearing gloves and other personal protective clothing, and to ease of decontamination.

Finally, ease of servicing should be evaluated. Batteries should be readily accessible without removal of other components or a large number of screws; this factor alone can save several maintenance man-hours per year. Plug-in circuit boards, accessible from both sides and with labeled components can also provide significant dollar savings in the maintenance areas. Circuit boards should also be keyed to prevent inadvertent erroneous positioning in the socket. This feature alone can sometimes save hours of trouble shooting, frayed nerves, and possibly components.

3.0. ELECTRONIC EVALUATION

Electronic evaluation is far more objective than the mechanical evaluation. Ten specific areas are evaluated; each is discussed in turn, and a specific testing procedure is cited.

3.1. POWER SUPPLIES

Most portable instruments use batteries as the prime energy source. The batteries may supply the power directly to the circuit, or in instruments of recent design, the batteries furnish energy to solid state supplies which convert the low voltage of several low cost, dry batteries to the high and intermittent voltages needed by the circuit.

3.1.1. Stability

Some power supplies have variable output, especially for the high voltage. With constant input voltage the output range is checked at the end points, and the stability noted. The measurements should include both voltage and current outputs. Generally, the circuitry of the instrument is used as the load, but fixed resistors can be used to provide loads of any desired value. The stability of the voltage and current output is observed as a function of load, output voltage, and input voltage. Modern power supplies should generally be capable of providing stable high voltages to within a few percent.

3.1.2. Temperature Dependence

Portable instruments are often used under field conditions with wide variations in temperature. Diurnal temperature variations of 50° F can be encountered, necessitating correction of instrument readings in many cases. Evaluation of the temperature dependence over the range of 0-120° is usually adequate, although extension of the extremes may be required.

Temperature dependence is determined by measuring the output over the desired range utilizing a constant input voltage. Generally, temperature effects are linear, and should be quite small. Battery voltage and current should also be recorded, for these may be limiting on the power supply.

3.1.3. Battery Life

For battery operated supplies, the output voltage and current are measured as a function of battery voltage. This test is best performed using the specific batteries designated for use in the instrument, since the proper operation of many power supplies is highly dependent on both battery voltage and resistance.

An instrument may be designed to operate with several types of primary or secondary batteries. Thus ordinary zinc manganese cells may be used for temperatures down to 0° F to 10° F. Below these temperatures zinc alkaline primary cells should be used. Battery life tests should cover those types of batteries which will be installed for these environments in which the instrument will be used.

Fresh, tested batteries should be installed in the instrument, and provisions made for measuring battery voltage, battery current drain, power supply voltages, power supply currents, and overall instrument response. The detector element of the monitoring instrument is placed in a radiation field which should provide a mid-scale reading on the most frequently used range. The instrument is turned on and the above variables recorded over the life of the battery. The end point of the battery is taken as that voltage at which the instrument response has decreased 10% from the response obtained with a fresh battery.

A battery test position is a highly desirable feature of any portable survey instrument. The battery voltage should be checked against the indicated position to ensure that it is properly located.

3.2. INPUT SENSITIVITIES

Most portable instruments are of the pulse type meter and aural output. The input sensitivity of voltage sensitive instruments (e.g., Geiger counters) is determined with a signal generator providing a voltage pulse similar to that given by the detector. The input sensitivity is expressed as volts or millivolts necessary to give a steady meter reading or aural output for a given pulse input repetition rate.

For current measuring instruments, such as ion chambers and proportional counters, a current source is used in lieu of the detector. The current necessary for full scale deflection is determined, and the input sensitivity expressed as the current for full scale deflection. If the active volume of the detector is known, the input sensitivity may be calculated in terms of R/hr required for full scale deflection. This calculated value can be checked by placing the instrument in a radiation field of the appropriate strength and measuring the current output of the detector. This technique can also be used to determine the linearity of the detector alone. Response in pulsed fields, which is daily assuming greater importance, can also be determined from this technique.

3.3. LINEARITY

Linearity of a rate meter is an important feature. It is easily and quickly checked by putting a signal similar to that provided by the detector into the detector input of the circuit. The signal generator should be initially set to provide a pulse rate (in the case of pulse detectors) or current (in the case of current measuring circuits) equal to about half scale deflection. In the case of charge sensitive input system, a voltage to charge convertor is put on the output of the voltage signal generator.

Linearity should be checked at several points over the range of 10 to 100% of full scale. It is important to watch the errors near full scale meter reading since some systems tend to saturate in this region. Linearity should be expressed in terms of the percentage deviation of meter reading from the correct value at any point on the scale rather than the full scale reading. Thus, an instrument that read 11% of full scale when the input signal was such as to give a reading of 10% of full scale, has a 10% deviation. A statement of $\pm$ % is most useful to bracket linearity.

3.4. RESPONSE TO ELECTROMAGNETIC
INTERFERENCE

Electromagnetic radiation of many kinds can cause unusual effects and spurious response. The circuitry, rather than the detectors, is more commonly affected. In most cases, well designed circuits with good electrical shielding will not be affected by electromagnetic fields.

3.4.1. Magnetic Fields

The meter is usually affected more frequently than other components by magnetic fields. The effect of magnetic fields can usually best be determined by actual observation under the specific conditions encountered in the field, orienting the instrument along an x, y, and z axis with respect to the direction of the magnetic lines of force. Differences in instrument response will usually be attributable to the magnetic field. The effects of magnetic fields can sometimes be negated by wrapping the instrument case with mu metal.

3.4.2. AC Induced Fields and Transients

Transients caused by interrupting an AC circuit under load are some of the worst offenders. This is especially true of AC line operated instruments, but also holds for battery powered instruments operated close to unshielded AC lines. Of course, in the case of battery powered instruments, radiated radio frequency (rf) is the cause of erroneous readings. Electric typewriters and calculators are frequently found in areas where monitors are used.

The effect of AC transients and induced fields can be crudely checked by using an electric drill or similar device. Transients and AC induced fields should not cause deflections greater than 10% of the meter reading.

3.4.3. Response to Radiated Energy (Radio Frequency)

The most common sources of radiated energy that interfere with monitoring instruments are:

1. Induction type heaters

2. Radio frequency welders

3. Radio frequency generated by AC line transients

4. Radar beams

5. Ignition systems.

Welding and radar apparatus may have x-radiation associated with it, and care must be taken to ensure that the response of the instrument was not elicited by stray x-rays. Interference from ignition systems can be particularly vexing. Errors greater than $\pm$ 10% of a given measurement have been noted at distances of 200 feet from the source of interference.

In general, instruments should be field tested for radio frequency response under the conditions in which they will be used. The untoward effects of R.F. can sometimes be alleviated by a metallic case, properly grounded.

3.4.4. Electrostatic Charges

As a rule, ion chamber instruments are the ones most sensitive to static charges. The effect of electrostatic charge is easily checked qualitatively by moving a charged comb or piece of plastic around the instrument and detector, and noting any deviation in meter reading. The deviation should not exceed $\pm$ 10% of the reading with static charge; meter deflection from zero should be minimal—not more than 1% of full scale.

3.5.   SWITCHING TRANSIENTS

Switching transients refer to quick, violent, large excursions of the meter when the range switch is changed from one range to the next. Switching transients should not drive the meter pointer off scale ("peg" the meter). As a guideline, the pointer should return to the original zero reading in two seconds or less, without the aid of a meter reset switch.

When checking switching transients, the instrument should be allowed to warm up completely, and the range switch changed stepwise in both directions.

3.6.   CAPACITANCE EFFECTS

Capacitance effects are changes in meter readings or instrument response from physical motion of parts of the instrument. Historically, the word capacitance is used, for in the early days of radiation monitoring instruments, case movements would cause capacity changes in the circuit, which would lead to changes in meter readings. A second type of capacitance effect--hand capacitance--is noted when the hand is brought near the case or detector.

In general, capacitance effects shall not cause meter deflections in excess of $\pm$ one minor meter scale division. To check for capacitance effects, the instrument is turned on and allowed to stabilize on the most sensitive range. Pressing on any part of the case, control knobs, lifting the control knobs lightly or any other manipulation which is associated with normal procedure shall not cause meter deflection in excess of that listed above. The test should be repeated on all ranges.

3.7. GEOTROPIC EFFECTS

Geotropic or position dependence effects are related to capacitance effects. In an ideal instrument the meter reading will not change regardless of the position in which the instrument is operated. Generally there is some position dependence due to gravity effects on the meter.

Position dependence can be quickly and easily determined by rotating a stabilized instrument through a full 360° in two planes normal to each other and parallel to the ground. Ordinarily, deviations of less than 1 or 2% of the full scale reading are acceptable.

3.8. TEMPERATURE DEPENDENCE

The effects of temperature on the total functional instrument package is of prime importance, since instruments may be used in thermal environments ranging from sub-freezing to in excess of 120°F. It is desirable that the temperature coefficient of the instrument be less than $\pm 0.1\%/°F$. Generally, the detector is the major contributor to temperature dependence, but other components may also contribute significantly. Occasionally, a single component such as a resistor may require replacement in order to minimize temperature effects.

Evaluation of temperature dependence requires an environmental or temperature test chamber. The instrument, detector, and source to give a constant meter reading are placed in the test chamber. The temperature is raised at a rate of about 20° F/hr until a maximum temperature of 140° F is reached. Meter readings are taken about every 10 degrees F, and a plot made of scale reading vs. temperature made. The temperature is then lowered at the rate of 20° F/ hr until the lower limit of -10 degrees F is reached. Meter readings are taken every 10 degrees F. When the lowest temperature has been reached, the temperature is again raised at the rate of about 20° F/hr until room temperature is reached. As previously, readings are taken about every 10° F.

The test at lower temperatures is also a test for high humidity operation. At the start of the test, the air is at room temperature in the test chamber, and normally has a relative humidity of 30 to 50%. As the temperature is lowered, the air in the chamber will become saturated and moisture may form inside the instrument case. In most instances trouble is not encountered with moisture until the temperature is increased from the lowest point back towards room temperature.

In the above temperature tests there may be times when large calibration deviations occur. It will then be necessary to check sub-units of the system one or two at a time, until the temperature sensitive parts are determined.

3. 9.   EXTRACAMERAL EFFECTS

Extracameral effects are meter deflections caused by interaction of the ionizing radiation with part of the instrument other than the true detector or ionization chamber component. The effect is most prevalent in ionization chamber type instruments.

Obviously, it is necessary to check extracameral effects by shielding the detector or limiting the size of the radiation beam. By using a well collimated beam, 2mm wide, a scan can be made across the instrument. The meter reading can then be plotted as a function of location. In general, extracameral response should be less than a few percent of the average reading obtained with the beam directed through the detector, regardless of range. As a rule, the extracameral effect will decrease proportionately as the instrument is switched to higher ranges.

3.10. SOUND AND VIBRATION EFFECTS

Response to sound occurs mostly in pulse type instruments. At times the speaker response may cause unwanted feedback, negating the use of the instrument. Ideally, noise effects should be checked in a sound chamber with variable intensity pure tones. The test is made by exposing the instrument to noises of various frequencies and intensity, and any abnormal response is noted. This is obviously impractical, and so random noise generators or general industrial noise sources can be used. Response to vibration or impact noise may result in temporary or permanent change in calibration. In addition, vibration may cause physical damage to the instrument. A shaker table is a good device for providing vibration for test purposes. One of the most satisfactory and practical vibration tests is to field test the instrument for a period of time. Ordinarily, noise and vibration should not noticeably affect instrument performance.

4.0. RADIOLOGICAL EVALUATION

The radiological evaluation is perhaps the most germane and certainly the most meaningful to the operational health physicist. It is also the most objective, and requires an extensive amount of evaluation equipment. In general, the radiological evaluation is limited to the detector, or the response of the instrument as a whole.

Characterization of instrument response is rendered difficult by the problem, and one not unique to instrument evaluation, of terminology. In the succeeding paragraphs, as in the above sections, an attempt will be made to define or limit terms, in the hope of improving communication.

4.1. RANGE

The term range covers a multitude of features. The range is, of course, the whole of the instrument response. For instruments with linear readout, the range is ordinarily expressed in terms of zero to maximum full scale reading obtainable. For instruments with logarithmic readout, the range must necessarily be expressed in terms of the minimum and maximum scale readings obtainable.

In the typical evaluation, the range of the instrument is usually not specifically checked, for other aspects of the evaluation will adequately cover this area. However, a rapid check of the range can be easily made by simply varying the source to detector distance.

4.2. SENSITIVITY

The sensitivity of an instrument describes its capability in discriminating between two approximately equal quantities. Sensitivity should be constant over the entire instrument range, and is best expressed in terms of percent of full scale for instruments with linear readout. For instruments with logarithmic readout, sensitivity should be expressed as percent of the actual scale reading with the scale reading stated.

To illustrate, a portable instrument with linear response has a range of 0-10 mR/hr on a single scale. It is possible to read the meter to the nearest 0.5 mR/hr. The sensitivity of this instrument is $0.5/10 \times 100 = 5\%$. Similarly, an instrument logarithmic response has a range of 0.1 to 10 mR/hr. Near the low end of the scale, changes of 0.02 mR/hr can be determined, rising to 2 mR/hr at the high end. In this case, the sensitivity is 20% at both ends of the scale. If the minimum detectable changes were 5 mR at the high end, the sensitivity would be expressed as 20% at 0.1 mR/hr to 50% at 10 mR/hr. An expression of $\leq 50\%$ for the sensitivity, although valid, would not be as descriptive as the previous method.

Sensitivity is determined with the instrument in an appropriate radiation field, utilizing a device such as a trolley or carriage to permit small and reproducible changes in the source to detector distance. In this manner, small yet accurate variations in dose rates can be obtained. Sensitivity should be determined at at least two points—one each near the maximum and minimum scale reading.

In the case of alpha monitoring instrumentation, sensitivity is expressed in the manner described above, but is determined with several sources, each having slightly different strength. A single source covered with mesh to absorb various fractions of the alphas, will suffice.

4.2.1. Detection Limit

The detection limit is the minimum (or, less commonly, the maximum) radiation reading which can be obtained, with the instrument. This value can be obtained from the sensitivity.

4.3. ACCURACY

Accuracy is, of course, the relationship of the instrument reading to the true value, and is expressed as a limit or range in terms of percent of the true value. Accuracy is the sum total of all factors which may adversely affect the reading of the instrument, including reproducibility (4.4), stability (4.10), and sensitivity (4.2).

Although correctly, accuracy is a statistical phenomenon, calculable from the known variance of the numerous factors that affect the reading of the instrument, such calculation (and indeed even securing the data on which to base the calculation) is impractical. Hence, the term accuracy generally refers to the total uncertainty ($\pm$ 3 standard deviations) in the instrument reading, with the measurement made under nearly ideal conditions, and with appropriate corrections made for such factors as temperature, pressure, energy, and geometry.

Accuracy is best determined with a calibrated source. Instrument readings over the entire range are compared with the "true" value as determined from the source strength. Presented on the following page are some actual data obtained with an ion chamber survey meter:

True mR/hr	Instrument mR/hr	Instrument Range
15	14	× 1
30	30	× 1
45	47	× 1
50	45	× 10
100	95	× 10
300	310	× 10
450	460	× 10
1500	1600	× 100
3000	3000	× 100
4500	4800	× 100

This instrument was rated as having an accuracy of $\pm 10\%$.

4.4. REPRODUCIBILITY

Reproducibility, or precision, is a measure of consistency of readings. It is determined by making at least five readings in a given field, removing the field between each reading. This should be done over at least three points on the range of the instrument. The following data were obtained with a GM survey meter on the × 10 range:

"True" CPM	Observed CPM
1,000	1,050; 1,050; 1,100; 1,050; 1,100
3,000	2,900; 3,000; 2,950; 2,900; 3,000
5,000	5,100; 4,900; 4,950; 5,200; 5,000

Reproducibility of better than $\pm$ 5% is indicated by the data.

Note that reproducibility is not the same as accuracy; an instrument that always reads exactly 1.50 the true value would be 100% reproducible, but would have an accuracy of +50%.

4.5.   SATURATION

When in radiation fields above their intended range, some instruments will saturate. The effect manifests itself in one of two ways: either the instrument reading rises to a fraction—usually about 80-90%—of the maximum full scale reading and remains there regardless of how intense the radiation level gets, or, in the instrument is driven off scale but the reading drops back down to zero as the field intensity increases. Each of these conditions can result in a serious situation with possible overexposures to personnel. The former effect occurs most commonly in ion chamber instruments, and the latter in Geiger type survey meters.

Saturation is easily checked by placing the instrument in an appropriate radiation flux, and increasing the intensity until the effect is noted or a predetermined level—usually 100 times the maximum range of the instrument—is reached.

4.6.   ENERGY DEPENDENCE

Energy dependence--also known as spectral sensitivity--refers primarily to the response of an instrument to x or gamma radiation of different energies. Energy dependence is caused by many factors, the two most important ones being photoelectrons (and, to a lesser degree, Compton electrons) from the detector wall, and self-absorption within the detector wall. These effects are, of course, competitive.

To evaluate energy dependence, the detector is placed in a field of known strength and energy and the response compared with the true exposure rate. Ordinarily, the instrument is calibrated with a source having an energy of approximately 1 MeV, and the instrument response normalized to this energy.

Several effective energies in the range of 10 keV to greater than 1 MeV are necessary to achieve an accurate indication of energy dependence. An x-ray machine--preferably with 300 kVp capability--can be used to obtain effective energies in the 50-250 keV$_{Eff}$ region by means of heavily filtered spectra; a wide beam K fluorescent source can be used to conveniently obtain essentially monoenergetic photons in the region 8-100 keV$_{Eff}$. The higher energies are provided with nuclides, commonly ^{137}Cs (662 keV$_{Eff}$), ^{222}Ra + daughters ($\sim$ 800 keV$_{Eff}$), and ^{60}Co (1.25 MeV$_{Eff}$). A typical series of energies for a cutie pie type instrument might be 8, 17, 23, 40, 60, 80, 100, 125, 175, 200, 662, 1250 keV$_{Eff}$.

Energy dependence should be determined with both open and closed window, and often for other conditions, such as with the beam end--on or through a thicker side wall. Energy dependence should be

reported as a +, - percentage over a specific energy
range or a +, - percentage normalized to a specific
energy. Graphical presentation, in the form of a plot
of response per unit exposure as a function of energy,
is a superior method of showing energy dependence.

4.7. TEMPERATURE AND PRESSURE DEPENDENCE

These effects are usually minor, and are commonly caused by changes in the mass of air within
the detector. However, temperature effects on other
components can also cause appreciable changes in
instrument response, and a testing procedure similar
to that described in 3.8 above is recommended.

Changes in atmospheric pressure should have
minimal effect. If an appropriate environmental test
chamber is available, the change in instrument response to a constant field should be determined in
10 mm increments over the range 720-780 mm Hg. A
pressure coefficient, in terms of percent change per
mm Hg will usually be apparent from the data.

4. 8. ANGULAR DEPENDENCE

Because most detectors are not spherical, angular dependence may cause serious discrepancies in
readings. In particular, angular dependence becomes
significant when a fixed geometry or windowed detector
is used in an ambient field or in such a manner that
the window cannot be pointed towards the direction of
the field.

Angular dependence is best checked through a
full 360° in perpendicular planes, one parallel to the

horizontal and the other to the vertical axis of the
detector. Measurements should be made at 15° increments. If the detector is symmetrical about an
axis, the determination can be made over 180° or
90°, as indicated by the geometry.

The statement of angular dependence should be
expressed as $\pm$ a percentage from a fixed point, usually the so-called "normal" of the detector. However,
a graphical representation is superior, as is true for
energy dependence.

4. 9. RESPONSE TO UNWANTED RADIATIONS

Portable survey meters are usually intended to
monitor one type of radiation. Photon monitors such
as ionization chambers, for example, should be relatively insensitive to other penetrating radiations, viz.
neutrons. Similarly, alpha or neutron monitors
should be insensitive to photon radiations.

Checking the rejection of unwanted radiations is
often more art than science. For example, energy
dependence must be considered, particularly when
evaluating the rejection of photons by neutron or alpha
scintillation instruments. Neutrons may show a similar energy dependence, with a response being elicited,
say, only by thermal neutrons. And, the undesired
response may not appear except in a mixed field.
However, as a general rule of thumb, it is usually
necessary to check response to unwanted penetrating
radiation up to levels of about 10 R/hr in the case of
photons, 10^7 n/cm^2/sec of fast neutrons, and 10^8
n/cm^2/sec of slow or thermal neutrons.

5. 0. CONCLUDING REMARKS

The evaluation of portable radiation monitoring instruments is of prime importance if these devices are to be selected intelligently and used to maximum advantage by operating health physics personnel. Knowledge of appropriate correction factors, idiosyncrasies, and response characteristics are vital to proper application in the field.

Evaluation of a portable instrument should take into account relevant mechanical, electronic, and radiological response features, and should be oriented towards ultimate field applications. Both physical and psychological factors should be considered; personnel will not have confidence in an instrument that doesn't look as if it will work. Underlying all phases of the evaluation is economics; one must get the most for each dollar spent. Implicit in this is minimal maintenance along with low initial cost.

Proper evaluation requires time and a fair amount of expensive test equipment. The program outlined and briefly discussed above is by no means the ideal nor is it all encompassing. Rather it represents an attempt—perhaps more realistically, a start—towards uniform evaluation procedures and terminology, with a practical end result.

PLUTONIUM SURVEY WITH AN X-RAY
SENSITIVE DETECTOR*

J. F. Tinney, J. J. Koch and C. T. Schmidt
Lawrence Radiation Laboratory,
University of California
Livermore, California

ABSTRACT

Traditionally, plutonium surveys have been conducted with alpha-sensitive detectors. More recently,
however, x-ray detectors have been developed for this
purpose. This report describes the design and performance of the low-energy x- and gamma-ray detector developed at LRL for plutonium survey. Presented
also are the results of a plutonium survey conducted
under field conditions with this detector and a conventional alpha survey instrument.

INTRODUCTION

The widespread use of plutonium and the attendant
possibility of accidental release have created the need

---

*Work performed under the auspices of the U. S.
Atomic Energy Commission.

851

for an instrumentation system that can delineate rapidly and accurately large contaminated areas under adverse field conditions. Traditionally, plutonium surveys have been conducted with alpha-sensitive detectors. More recently, however, x-ray detectors have been developed for this purpose.

The advantages of x-ray detection for plutonium survey include the following:

1. The detector need not be placed in contact with the contaminated surface. Thus, detector damage is minimized and the need for point-by-point measurements is eliminated.

2. Detector performance is less seriously degraded by a thin overburden of weeds, water, snow, dust, etc.

3. The survey can be conducted after the radioactivity has been "tied down" and the hazard minimized.

This paper describes the design and performance of the FIDLER* x-ray detector being developed at LRL for plutonium survey. Presented also are the results of a plutonium survey conducted under field conditions with FIDLER and a conventional alpha survey instrument.

RADIATION DATA

Isotopic mixtures of plutonium emit uranium and neptunium L x-rays (13.6, 17.2, and 20.2 keV) that

---

*Field instrument for detecting low energy radiation.

are produced by internal gamma-ray conversion in the daughters of plutonium and [241]Am, respectively. The [241]Am, which also emits 60-keV gamma rays, is formed through the beta decay of [241]Pu. As a result of these processes, the x- and gamma-ray specific activities of most plutonium samples will depend on initial isotopic composition and age.

Table 1 shows a tabulation of pertinent radiation data for several plutonium isotopes and [241]Am. The

Table 1.

Radiation data for several plutonium isotopes and [241]Am.

Isotope	Alpha specific activity (dis/min/ μg)	X rays/ alpha	X-ray equivalence (μg Pu/ μCi [241]Am)
[238]Pu	3.85×10^7	0.1055	2.05×10^{-1}
[239]Pu	1.36×10^5	.048	1.27×10^2
[240]Pu	5.02×10^5	.10	1.59×10^1
[241]Pu	2.49×10^8	–	–
[242]Pu	8.67×10^3	.10	9.63×10^2
[241]Am	7.14×10^6	0.376	–

influence of age on isotopic composition will be discussed in a subsequent section of this report.

DETECTOR

The FIDLER low-energy x-ray detector consists of a 5-in.-diam by 1/16-in.-thick NaI (T1) crystal* optically coupled through a quartz light pipe to a selected RCA8055 multiplier phototube. The entrance

---

*Harshaw Chemical Company, Cleveland, Ohio.

window is 0.010-in. beryllium, and the entire assembly is housed in a 5/32-in. stainless steel can.

The resolution of the detector for low-energy photons is illustrated in Fig. 1. This spectrum was obtained with a 10.7-μCi ^{241}Am source at a source-to-detector distance of 30.5 cm. The detector output was recorded with a Nuclear Data 2200 multichannel analyzer and the ancillary electronic equipment which included a high-voltage power supply, low-noise preamplifier, and a linear amplifier.

Also shown in Fig. 1 is the background spectrum that was recorded in the laboratory. The energy bands, bracketing the 17- and 60-keV photopeaks, indicate the optimum discrimination levels for integral counting. These levels were calculated on the basis of obtaining a maximum signal/background counting rate ratio.

At 60 keV, the detector resolution is 19.2 percent FWHM* and the resolution for the unresolved 17-keV x-ray band is about 57 percent FWHM.

FIELD ELECTRONICS

Two commercially available single-channel analyzers have been tested with the FIDLER detector. Both of these units are battery operated and have integral count rate meters and regulated high voltage power supplies.

The analyzer window discriminators are adjusted to accept pulses in a preselected energy interval which is established from multichannel analyzer data to provide a maximum signal^2-to-background count-rate ratio.

---

*Full width at half maximum

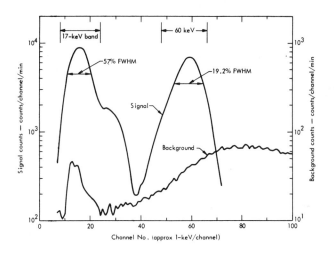

Fig. 1. Spectra for 241Am and background as measured with a 5-in. -diam by 1/16-in. -thick Nal (T1) scintillation detector. Source was 10.7-μCi 241Am, and source-to-detector distance was 30.5 cm. Optimum discriminator settings for 17-keV band and 60 keV are shown.

CALIBRATION AND PERFORMANCE

To establish the relationship between detector counting rate and plutonium contamination level, a number of factors must be given careful consideration. These include the following:

1. The isotopic composition of the plutonium contaminant.

2. The thickness and composition of any overburden material.

3. The influence of other radioactive materials that may be present.

4. The size of the contaminated area.

5. The characteristics of the radiation detector.

6. The source-to-detector geometry.

The relatively simple calibration procedure described in this section has been developed to examine the influence of these parameters on plutonium detection sensitivity for the FIDLER.

Detector Calibration

There are several methods of calibrating a portable survey instrument so that detector counting rate can be related to surface contamination level. The most obvious approach is to suspend the detector above a uniformly contaminated surface of known activity. This method was used to evaluate the performance of the prototype FIDLER detector, but the sources required careful handling and could not be used in the field. A somewhat less complicated calibration procedure based on the use of a single point source was therefore developed.

This source-to-detector geometry for the simplified calibration procedure is illustrated in Fig. 2. Here the detector counting rate is recorded at a number of source locations as a point source is moved radially away from the vertical axis of the detector. The results of these measurements can then be used to calculate area source detector sensitivity as a function of the radius of the contaminated area.

As an example of how this is accomplished, we can start with the count rate versus radial displacement data presented in Fig. 3. These count rates were recorded using a 10.7-μCi ^{241}Am source (17-keV Np L x-ray band) and a detector height of 30.5 cm. For a source-to-detector geometry with cylindrical symmetry, a point source at any given radial displacement will produce the same detector counting rate as a narrow annular ring source having the same radius and total activity. We can therefore consider an area source as being composed of a number of uniformly contaminated concentric annular rings, as illustrated in Fig. 4.

If the counting rate as a function of radial source displacement can be written in equation form, the area source sensitivity can be obtained directly by integration as follows:

$$S_A = \frac{1}{Q} \int_0^R C.R.(r) \, 2\pi r dr \qquad (1)$$

where

$\quad$ S_A = detection sensitivity (counts/min/μCi/m^2),
$\quad$ Q = source activity (μCi),
$\quad$ R = radius of contaminated area (meters), and

J. F. TINNEY ET AL

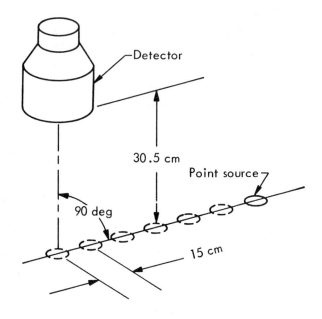

Fig. 2.   Geometry Used for Point Source Calibration Procedure.

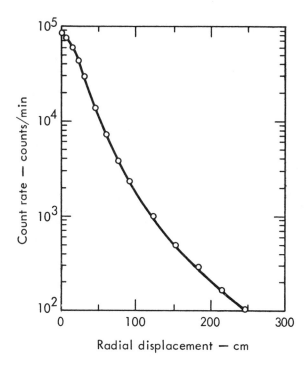

Fig. 3. Detector count rate (17-keV x-ray band) as a function of point source radial displacement. Source was 10.7-μCi ^{241}Am, and source-to-detector distance was 30.5 cm.

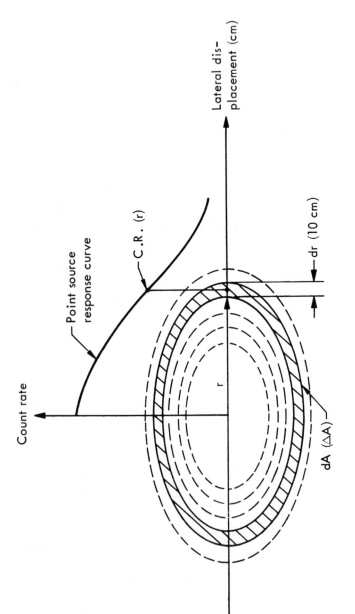

Fig. 4.   Pictorial Description of Integration Method.

C. R. (r) = detector counting rate for a source at
           radial displacement r (counts/min).

However, we can avoid the complication of deriving an
equation for C. R. (r) by using the approximate method
of integration described by the equation

$$S_A = \frac{1}{Q} \sum_r \text{C. R. (r)} \cdot \Delta A$$

$$= \frac{(2 \times 10^{-3})\pi}{Q} \sum_{r=5}^{r=R-5} \text{C. R. (r)} \cdot r \tag{2}$$

where

$r = 5, \; 15, \; 25, \; 35$ cm, and

ΔA = the incremental annular ring area.

Here we have considered the source as being composed
of concentric annular rings, each having a width of
10 cm (dr). Figure 5 shows the calculated area source
detection sensitivity for ^{241}Am based on Eq. (2) and
the data presented in Fig. 3. For large contaminated
areas, the detection sensitivity is fairly constant with
a value of about 3.85×10^3 counts/min/μCi/m^2 at
25 m^2.

Detection Limit Calculations

If a count rate meter is used to record the detector
output, the minimum detectable ^{241}Am point source,
at the 95 percent confidence level, can be calculated
from the following equation:

$$\text{MDPS} = \frac{3.3}{S_p} \sqrt{\frac{B}{2RC}} \tag{3}$$

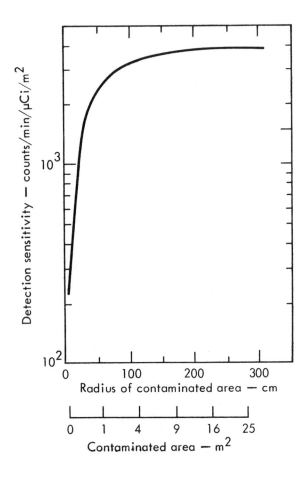

Fig. 5. Dependence of detection sensitivity on radius of contaminated area for [241]Am (17-keV x-ray band). Detector height was 30.5 cm.

where

MDPS = minimum detectable point source (μCi),

3.3 = number of standard deviations above background counting rate for detectability,

S_p = point source detection sensitivity (counts/min/μCi),

B = background counting rate (counts/min), and

RC = count rate meter time constant (min).

From the data in Fig. 3, S_p is about 7.85×10^3 counts/min/μCi for ^{241}Am (17-keV x-ray band). A similar equation can be used to calculate the minimum detectable ^{241}Am area source:

$$\text{MDAS} = \frac{3.3}{S_A} \sqrt{\frac{B}{2RC}} \qquad (4)$$

where

MDAS = minimum detectable area source (μCi/m^2), and

S_A = area source detection sensitivity (counts/min/μCi/m^2).

Since the background counting rate will not normally be known, it must be established from measurements outside of the contaminated area over similar terrain. Typically, the background counting rate for the 17-keV x-ray band will range from about 200 counts/min to 600 counts/min, depending on the type of soil and the geographical area. If we assume a background of 200 counts/min and an RC time constant of 12 sec (0. 2 min), the minimum detectable ^{241}Am point source is $9.41 \times 10^{-3} \mu$Ci. The minimum detectable area source is about $1.92 \times 10^{-2} \mu$Ci/m^2 based on the data presented in Fig. 5. Similar calibration procedures can be used to estimate detector response for a variety of low energy x- and gamma-ray emitting isotopes.

Detection Sensitivity for Various Plutonium Isotopes

Although detector calibration could be effected with a point source of each plutonium isotope using the method described above, isotopically pure sources of plutonium are difficult to obtain. The ^{241}Am data presented in Figs. 3 and 5, and measured L x-ray yields for the various plutonium isotopes (see Table 1) have, therefore, been used to establish point source and area source detection sensitivities for plutonium. This was accomplished by determining the quantity (μg) of each plutonium isotope that would provide the same 17-keV band x-ray activity as 1 μCi (alpha) of ^{241}Am. The ^{241}Am values for S_p, S_A, MDPS, and MDAS were then multiplied by the appropriate correction factors for each plutonium isotope. Results of these calculations are presented in Table 2.

Detection Sensitivity for Isotopic Mixtures of Plutonium

In practice, most plutonium samples will consist of a mixture of plutonium isotopes (238 through 242) as well as ^{241}Am resulting from the beta decay of

Table 2.

Calculated point and area source response data for various plutonium isotopes.

Isotope	S_p (counts/min/μg)	S_A (counts/min/μg/m^2)	MDPS (μg)	25 m^2 area MDAS (μg/m^2)
^{238}Pu	3.83×10^4	1.88×10^4	1.93×10^{-3}	3.94×10^{-3}
^{239}Pu	6.18×10^1	3.03×10^1	1.20×10^0	2.44×10^0
^{240}Pu	4.94×10^2	2.42×10^2	1.50×10^{-1}	3.05×10^{-1}
^{241}Pu	–	–	–	–
^{242}Pu	8.16×10^0	4.0×10^0	9.06×10^0	1.85×10^1

^{241}Pu. The exact isotopic composition will depend on the irradiation history, chemical separation, and age of the material.

The x-ray (17-keV band) specific activity of a mixture of plutonium isotopes can be calculated from the following equation:

x-ray specific activity (x-rays/min/μg mixture)

$$= W_{38}A_{38}R_{38} + W_{39}A_{39}R_{39} + W_{40}A_{40}R_{40} \qquad (5)$$
$$+ 387 \, W_{41} \, t + W_{42}A_{42}R_{42}$$

where

W_{38} = weight fraction of ^{238}Pu in mixture at time of chemical separation,

A_{38} = alpha specific activity of ^{238}Pu (α/min/μg), and

R_{38} = x-ray-to-alpha ratio for ^{238}Pu etc., and

t = the time from chemical separation (days).

The term $387 \, W_{41} \, t$ describes the increase in ^{237}Np L x-ray intensity resulting from an initial ^{241}Pu weight fraction, W_{41} ($t \leq 5000$ days).

Based on the x-ray specific activity calculated from Eq. (5) (or obtained from laboratory sample analysis), we can determine the quantity (μg) of plutonium mixture that would yield the same 17-keV x-ray activity as 1 μCi (alpha) of ^{241}Am. The ^{241}Am can then be used as outlined in the previous section to calculate values of S_A, S_p, MDPS, and MDAS.

Under certain conditions (i.e., aged plutonium and/or substantial overburden), the 60-keV gamma rays from ^{241}Am will provide better detection sensitivity than the 17-keV x-rays. Calibration at this energy was therefore obtained with a 10.7-μCi ^{241}Am source and a detector height of 30.5 cm using the geometry depicted in Fig. 2. The results of these measurements are presented in Figs. 6 and 7. The calculated values of S_p, S_A, MDPS, and MDAS for ^{241}Am, based on the laboratory background of 600 counts/min (see Fig. 1) are S_p = 6.73 × 10^3 counts/min/μCi; S_A = 3.55 × 10^3 counts/min/μCi/m^2; MDPS = 1.90 × 10^{-2} μCi; and MDAS = 3.6 × 10^{-2} μCi/m^2.

If the 60-keV ^{241}Am gamma-ray intensity is measured during a plutonium survey, the gamma-ray specific activity of the contaminant must be known. This can be established from sample analysis in the laboratory, or calculated from the following equation:

gamma-ray specific activity (γ/min/μg mixture)

$$= 387 \, W_{41} \, t \tag{6}$$

Unfortunately, the use of this equation requires a knowledge of the initial ^{241}Pu weight fraction and the age of the material; data that may be difficult to obtain.

After the gamma-ray specific activity has been established, the quantity (μg) of plutonium mixture

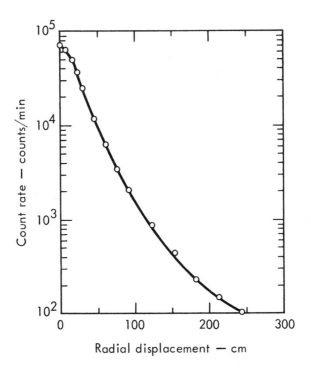

Fig. 6. Detector count rate (60-keV gamma ray) as a function of point source radial displacement. Source was 10.7-μCi ^{241}Am, and source-to-detector distance was 30.5 cm.

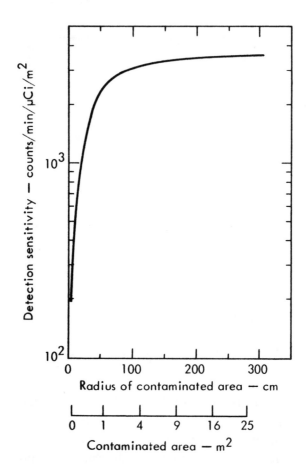

Fig. 7. Dependence of Detection Sensitivity on Radius of Contaminated Area for ^{241}Am (60-keV X-ray Band). Detector Height was 30.5 cm.

that would yield the same 60-keV activity as 1 μCi (alpha) of ^{241}Am can be calculated, and the ^{241}Am data used to determine values of S_A, S_p, MDPS, and MDAS.

FIELD TEST

The FIDLER instrument was recently used for plutonium survey at NTS. The survey was conducted over relatively flat terrain sparcely covered with creosote bush and sage brush. Earlier measurements with a conventional alpha survey instrument indicated that the contaminant was distributed over an area of approximately 15,000 m^2.

Survey Procedure

The detector was carried over the contaminated surface at a height of 27 cm and count rate meter readings were recorded at the locations indicated in Fig. 8. For comparison, readings were also taken at these locations with an alpha survey instrument, a Geiger-Mueller detector, and with the FIDLER detector at a height of 100 cm. Calibration was checked periodically with an ^{241}Am source and remained unchanged so that recalibration in the field was not required. Temperature variations during the survey were less than $\pm 10°$ F.

Results

It was immediately apparent from the instrument readings that the contaminated area contained a number of easily identifiable point sources. In addition,

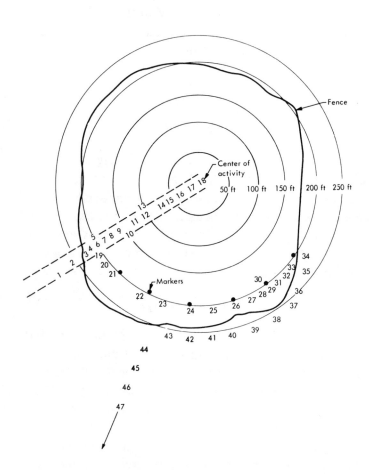

Fig. 8. Plutonium Survey Map (NTS).

several large areas were located over which the detector counting rate was nearly constant. Presumably
the contaminant in these areas consisted of a fairly
uniform distribution of small sources or radioactive
dust. By moving the detector in an "elephant trunk"
fashion, the boundaries of these areas were readily
delineated.

Count rate readings for the three instruments at
the locations indicated in Fig. 8 are presented in
Table 3. Although the interpretation of these measurements is complicated by a lack of information concerning the isotopic composition, age, and variations
in depth of burial of the plutonium contaminant, it is
interesting to compare the counting rates observed
with the alpha and x-ray detectors. In Fig. 9, FIDLER-
to-alpha-detector count rate ratios are presented for
four different types of contamination: general area,
wood chips and boards, metal chips, and wires. These
results show that the FIDLER had a higher counting
rate than the alpha detector for large contaminated
areas. For point sources this situation was reversed,
with the exception of one point source which exhibited
a very low alpha activity (No. 31). It should be noted,
however, that the alpha measurements could only be
made after the point sources had been located with the
FIDLER which had an effective range of 3 to 5 ft under
these conditions.

Measurements outside the contaminated area but
over similar terrain indicated a background counting
rate (17-keV x-ray band) of about 400 counts/min.
This would correspond to a minimum detectable ^{239}Pu
contamination density of 4 to 5 μg/m^2.

Table 3.

Comparative field survey.

No.	Description	Alpha survey meter, contact (α counts/min)	Geiger-Mueller tube, window open (mR/hr)	FIDLER, 27 cm (gamma counts/min)	FIDLER, 100 cm (gamma counts/min)
1	General area	5,000		7,500	2,200
2	General area	3,750		4,000	2,000
3	1-cm² metal chips	350,000		44,000	4,000
4	10-cm-long wire	300,000		12,000	3,000
5	General area	2,000		8,000	3,000
6	8-cm-long wire	190,000		52,000	8,000
7	General area	200	Negligible	20,000	13,000
8	40-cm-long wire	Negligible	0.4	49,000	9,000
9	20-cm-long wire	250,000	0.15	37,000	19,000
10	General area	10,000	0.1	51,000	36,000
11	20-cm-long wire	>500,000	0.5	130,000	23,000
12	General area	2,500	Negligible	20,000	20,000
13	General area	20,000	0.15 at 1 meter / 0.3 at contact	110,000	90,000
14	Wood (board)	50,000	0.3 at 1 meter / 0.4 at contact	220,000	160,000
15	Wood (board)	100,000	0.7 at 1 meter / 1.0 at contact	>500,000	>500,000
16	General area			850	
17	4-cm² metal chips	>500,000	0.05 at 1 meter / 1.0 at contact	120,000	10,000
18	Wood (board)	30,000	Negligible	36,000	11,000
19	1-cm² metal chip	50,000	Negligible	24,000	9,000
20	General area	300	Negligible	7,000	9,500
21	General area	100	Negligible	5,000	8,000
22	General area	Negligible	Negligible	3,900	7,000

23	4-cm^2 metal chip	>500,000	0.06 at 1 meter 0.5 at contact	55,000	15,000
24	General area in 1-ft-deep × 8-ft-wide ditch			17,000	17,000
25	General area	300		29,000	19,000
26	Wood chips in 2-m^2 area	5,000-25,000		70,000	50,000
27	General area on 3.5-ft-high × 10-ft-wide berm	100	Negligible	5,500	12,000
28	Wood chips at base of berm	15,000	0.05 at 1 meter 0.06 at contact	140,000	32,000
29	General area	Negligible	Negligible	4,600	8,000
30	Chips of wood in bush	Can not read	0.06 at top of bush	80,000	26,000
31	4-cm^2 metal chip	2,000	Negligible	7,000	10,000
32	4-cm^2 metal chip	>500,000	0.05 at 1 meter 2.0 at contact	180,000	32,000
33	General area	<50	Negligible	4,800	8,000
34	Wood chips	40,000	0.06 at 1 meter 0.5 at contact	60,000	9,000
35	General area			26,000	
36	General area			4,600	
37	General area			4,000	8,000
38	General area			3,100	
39	General area			3,500	
40	General area			2,100	
41	General area			4,000	
42	General area			1,700	2,400
43	General area			1,200	
44	General area			1,000	
45	Bush			5,000	
46	Continuous reading			~400	

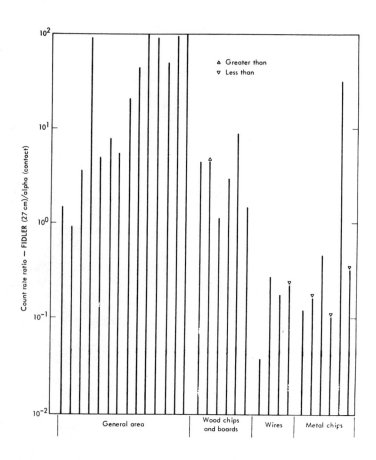

Fig. 9. Ratio of FIDLER-to-Alpha-Detector
Counting Rate for a Number of Samples.

SUMMARY

The results presented in this report indicate that a low-energy x-ray detector can be used to rapidly delineate large contaminated areas and localize "hot spots" of plutonium activity that would be difficult, or impossible, to detect with an alpha survey instrument.

The relatively simple calibration procedure that has been developed should facilitate the assessment of plutonium contamination levels under field conditions.

Additional experiments are presently being conducted to evaluate a new technique for estimating plutonium contamination density, based on intensity measurements in two different energy bands.

A MULTIPURPOSE
RADIATION MONITORING SYSTEM

F. E. Owen
Radiation Practices Engineering
Douglas United Nuclear, Inc.
Richland, Washington

INTRODUCTION

The multipurpose system features dual alarms;
an audible radiation emergency alarm that can be
activated by a very high dose rate and a visual radiation control alarm that is triggered by a nominal but
unplanned increase in working dose rates.   This
latter feature is of particular value at the Hanford
Production Reactors, operated for the AEC by Douglas
United Nuclear, Inc. These reactors have a large
graphite core, traversed horizontally by a few thou-
sand process tubes that hold the aluminum clad uranium fuel elements. Operators charge the fuel elements
into the front face of the reactor after they have removed the caps on the rear face where the irradiated
fuels are discharged.
The process tubes are also charged with other
aluminum clad elements of target material for isotope

production, dummies serving as spacers and neutron "poison" material to shape the flux.

Need for a multipurpose radiation monitoring system was first established as a result of operating experiences typified by a minor incident which occurred several years ago prior to DUN takeover. Operating personnel, on the front face elevator, were attending to last minute details in preparation for restarting the reactor after an outage. Several columns of "poison" elements were being charged into empty process tubes full of water under 20 feet of head. Three operators with their supervisor were charging this "poison" into the tubes by hand. This job required opening a ball valve on the front nozzle to the tube and inserting the elements one at a time against the water flow. One of the operators opened a valve, turned, picked up a "poison" element and when he turned back, found that:

(1) An element was protruding from the nozzle, and

(2) The radiation alarm bell began to ring

It must be understood that the "poison" elements were reused and therefore slightly radioactive. On this occasion, the carts filled with these elements when moved on the front work platform increased the background enough to trigger the alarm intermittently. Since it was a long and difficult task to change the alarm trip setting, the supervisor would look at the meter needle of the amplifier when the bell began to ring to assure that it did not exceed a reasonable rate; this practice was tolerated rather than condoned.

On the described occasion, however, the meter needle was pegged at full scale. The supervisor knew

that if the valve was not closed immediately, a large
number of very radioactive "poison" elements would
be flushed from the reactor core onto the work platform. Even though he could not be certain of the dose
rate his people were subjected to, he chose to have the
valve shut before leaving the work platform. Had the
tube contained, by error or design, irradiated fuel
elements, workers could have been exposed to exces-
sively high dose rates. On the other hand, the supervisor's actions actually "saved" personnel exposure.
Many workers would have received significant doses
disposing of the radioactive "poison" elements if the
tubeful had been allowed to flush out onto the work
platform. Review of dosimeters after-the-fact established that 10 mrem or less of whole body dose had
been received in this case. Thus, the need for a dual
alarm system became apparent. Such a system could
initiate:

(1) Corrective action for the circumstances just
    described, and

(2) Immediate evacuation should a fuel element
    come out of the reactor core and expose
    workers to excessive dose rates

SYSTEM DESCRIPTION

The heart of the multipurpose radiation monitoring system is the log picoammeter. The face of this
unit can be recognized by the meter readout and two
adjustable alarm control knobs.

The log picoammeter has a six decade range
from 1 mR/hr to 1000 R/hr providing wide coverage

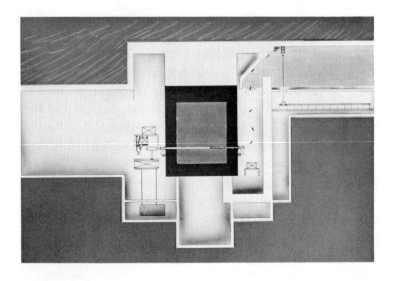

Fig. 1. Cut Away Diagram of a Hanford Production
Reactor

Fig. 2. Six Decade Log Picoammeter

without range change. The dose rates can be read within $\pm$ 10% over the full range of the meter which is considered a most reasonable degree of precision.

The two alarm trip controls operate independent of one another and of the meter readout. The high control can adjust the emergency alarm trip setting from 100 mR/hr to 1000 R/hr and operates a latching relay; the emergency alarm has to be silenced by a manual reset. This control knob is key locked providing a tamperproof emergency radiation alarm. The low alarm trip control, operating through a non-latching relay, is adjustable from 1 mR/hr to 10 R/hr.

The detection chambers are large air ionization chambers containing a small ^{137}Cs source that provides a background dose rate of 1 mR/hr. Thus, an equipment failure can be detected by the recorder when the readout drops to a zero dose rate. The chambers are located in trafficed Zones having a potential for a significant increase in gamma dose rate (i.e.: Zones facing the reactor top, bottom, sides, front and back faces as well as sites where radioactive sources can be removed from shielded locations). The chamber is positioned so it will detect a source as it emerges from a shielded location and as soon as it exposes any personnel present.

Rotating beacon ray lights with yellow lenses are placed so the beam will be easily seen by anyone who might be subjected to an unplanned increase in dose rate. These yellow beacon ray lights have been standardized as the radiation control alarms within Douglas United Nuclear, Inc., facilities. They are actuated by the low alarm trip. Radiation monitoring personnel set the alarm trips after they have established the working dose rates within the Zone. The rotating

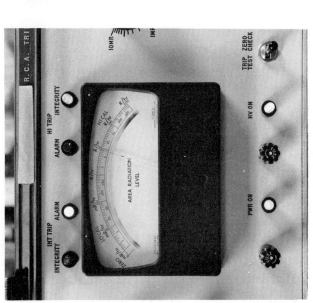

Fig. 3. Meter Face. The Needle Registers a 9 R/hr Dose Rate

Fig. 4. High and Low Alarm Trip Control Knobs

beacon ray light then is the means used to alert employees to an unplanned increase in their dose rate. The rotating yellow beam of light reflects off of walls and attracts the attention of everyone even when the Zone is well lighted. The workers then have 30 seconds to correct the "problem" or leave the Zone. Using light instead of sound, the radiation control alarm does not impair communications that are essential to implement corrective actions during the course of a minor incident.

Log picoammeters have been mounted with the other essential components of this multipurposed system for use on carts. The detection chambers are attached to 50 feet of cable to provide positioning flexability. During removal of a bucket of dummy element spacers from the storage basin, for example, the bucket can be closely monitored. And if the material is more radioactive than expected, the yellow light on the movable cart will flash until the operator at the hoist switch lowers the bucket back under water.

The radiation emergency alarm, a loud cyclic howler actuated by the high alarm trip, is located in the vicinity of the chamber. Response to this alarm is immediate and rapid evacuation from the Zone; dose rates are too high to allow time for thinking and/or corrective action. Impairment of voice communication under these circumstances does not constitute a problem, in fact, it may be desirable because the only proper action is immediate evacuation.

The idea underlying the radiation alarm settings for the dual alarm system is to identify the degree of loss of radiation control. Such identification assists the employee in minimizing his exposure to radiation without unnecessary work restrictions and in

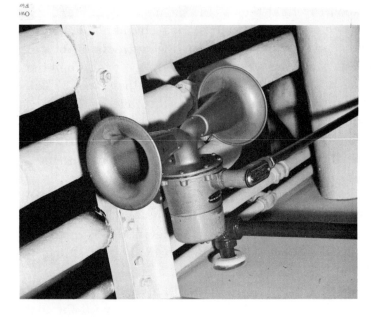

Fig. 6. Radiation Emergency Alarm

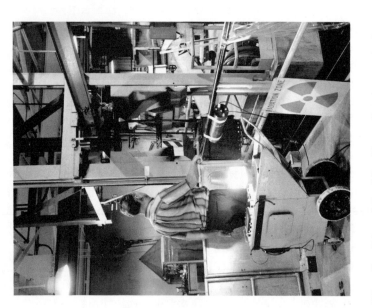

Fig. 5. Movable Monitor-Cartmounted
The Yellow Beacon Ray Light is Alarming

protecting against an accidental exposure in excess of
AEC standards. The rationale imbued in the system
includes:

1. When the radiation control alarm (yellow beacon ray light) activates, dose rates have increased 30% or more and the worker is to
   leave the Radiation Zone.

2. However if there is no emergency radiation
   alarm, the worker can be allowed up to 30
   seconds to correct an obvious condition without exceeding the AEC radiation standard of
   3 rem in a quarter whole body dose.

3. If the emergency radiation alarm has sounded,
   a worker must leave promptly and rapidly,
   otherwise he may receive a whole body dose
   exceeding the AEC standard.

4. Since the dose rate from a source is different
   for the chamber than for the worker, the radiation emergency alarm trip setting:

   a. Assumes a point source at a minimum dis-
      tance of 3 feet (about half a man's height)
      from the source to the worker, a practical
      consideration.

   b. Conservatively employs a dose rate rela-
      tionship in proportion to the inverse square
      of the distances.

   c. Assures that the system will alarm when
      the dose rate to a worker exceeds 360
      rems/hr (360 rem/hr equals 3 rem in 30
      seconds).

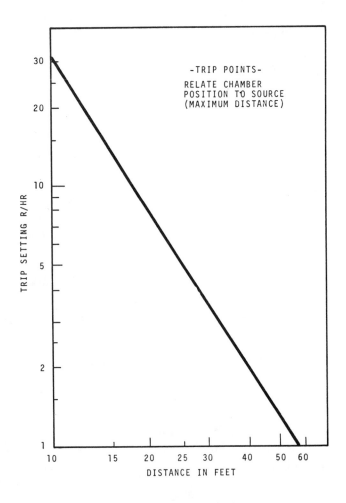

Fig. 7. Emergency Alarm Points

In each reactor control room there is a radiation control panel. This panel contains the dose rate recorders. Activated alarms and equipment failures anywhere in the facility are annunciated via a horn on top of the panel. Annunciation calls attention to the panel, and a light on the panel will indicate which alarm was triggered.

Because of a ^{137}Cs source in each detection chamber, a dose rate of 1 mR/hr is read on each recorder. A zero reading at the recorder shows that the measurement system is not measuring and/or relaying the chamber dose rate to the recorder. And such a reading will trigger the equipment failure alarm.

Near the top of the panel are two dual pen recorders that provide a continuous trace of monitors at four particularly important sites. Near the bottom of the panel is a single pen recorder that is used to obtain a continuous trace from a selected monitoring unit. The selector switches are located just below the recorder.

The 24 point main recorder in the center of the panel provides a complete environmental record for the reactor facility by recording points from all monitoring units including those feeding the continuous trace recorders. The pointer locks on the input signal from one monitoring unit for a few seconds while printing and then "moves on" to another.

Connected to the 24 point recorder is a pilot wire telemetry system that retransmits the radiation dose rates as received and recorded in the reactor control room to the Central Fire Station located some distance from the reactors.

F. E. OWEN

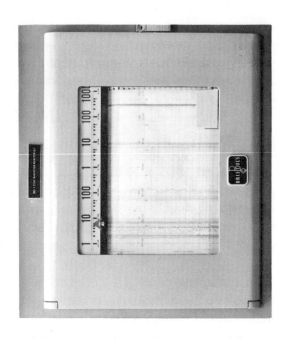

Fig. 9. 24 Point Recorder and Slide Wire Transmitter

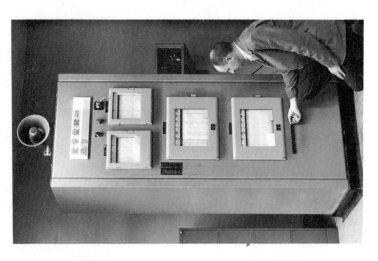

Fig. 8. Radiation Control Panel

The pilot wire telemetry system equipment is functionally divided into three sections: the telemetry console at the Central Fire Station, the transmitter units in the reactor facilities and the transmission lines. At the transmitter there is a 0-10 volt retransmitting slide wire that follows the 24 point recorder pen, feeding the FM (frequency modulated) solid state transformer that is capacity coupled to the system. The output of the transmitting unit is a FM carrier with the carrier frequency modulated by an amount proportional to the slide wire voltage. The output of each transmitter unit is fed over the fire alarm lines (i.e., typical telephone lines). Because the fire alarm signal is activated by a DC pulsed signal, there is no interference by the FM carrier signals from the radiation alarm-recording system.

The telemetry console at the Central Fire Station contains recorders, audible and visual annuciators, FM receiver and logic circuits capable of accepting signals from the transmitting units at the reactors. The console receiver remains on standby until actuated by a momentary loss of the FM carrier when a radiation emergency alarm is tripped by one of the monitoring units. An audible alarm alerts the fireman on duty, the visual display identifies the reactor facility and the recorder begins to operate. This single pen recorder provides a continuous trace of the 24 point recorder at the reactor. The chart moves about 3 inches a minute displaying the 24 points in a continuous line of steps as the recorder at the reactor hesitates at each point.

The fireman is on duty ready to contact emergency crews in case of fire or a serious radiation event.

F. E. OWEN

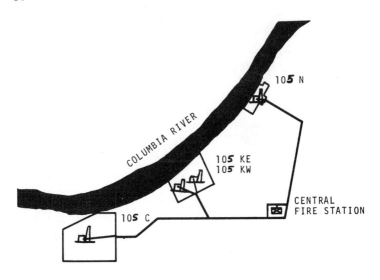

Fig. 10. Radiation Alarm--Recording Network

Fig. 12. Fire Station Control Room

Fig. 11. Telemetry Console

CONCLUSIONS

The multipurpose radiation monitoring system now in use at the Hanford Production Reactors by Douglas United Nuclear, Inc. , provides:

- An environmental radiation record on a single chart for each facility

- Control alarms to alert workers to changing dose rates and thus prevent unwarrented radiation exposure

- Radiation emergency alarms in readiness to warn workers of high dose rates and prevent exposure in excess of the AEC standards

- A system that is self-monitoring for equipment failures

- Readability over a wide range of dose rates from working levels to serious radiation event levels without instrument range changing

- Central readout and recording of the monitoring units and annunciation of alarms

- Remote annunciation, readout and recording by pilot wire telemetry in the unlikely event of a serious incident

ACKNOWLEDGMENTS

There were many contributors to the criteria, development, design, installation and revisions of this multipurpose system. Notable among the contributors were M. L. Smith, Health Physicist, McDonnell-Douglas; G. L. Erickson, Instrument Development Engineer, Douglas United Nuclear, Inc. and W. R. Thorson, Design Engineer, Douglas United Nuclear, Inc.

SOME USES OF THERMOLUMINESCENT DOSIMETRY AT ACCELERATORS*

Morris J. Engelke
Los Alamos Scientific Laboratory
University of California
Los Alamos, New Mexico

ABSTRACT

This paper describes some applications of thermoluminescent dosimetry to Health Physics problems encountered at accelerators. A convenient method of measuring integrated exposures in a mixed radiation field is discussed where miniature LiF dosimeters are placed in various diameter polyethylene spheres. Preliminary investigations using these spheres as a neutron spectrometer has been initiated. A rapid and simple assessment of personnel radiation exposure to gammas is described where miniature thermoluminescent dosimeters are incorporated into the frames of eyeglasses. The evaluation of the integrity of a shield wall is made by insertion of a 12-ft wand

*Work done under the auspices of the United States Atomic Energy Commission.

895

through the shield with TLD's spaced every four inches.
The practical use and limitations are discussed.

INTRODUCTION

Radiation levels at accelerators will vary greatly,
depending on operating conditions. When the beam is
"ON", the character of the radiation will be that of
neutrons and gammas, with intensities depending on
target material and beam operating parameters. With
the beam "OFF", the induced activities existing in accelerator structures present a wide spectrum of photon (and beta, even though the TLD doesn't differenti-
ate) energies and intensities. Personnel dosimetry in
the first condition involves mixed radiation fields that
occur outside of the biological shielding. In the second case, induced activities are proportional to the
average amount of power absorbed in the accelerator
structures and this may be a formidable problem.
Some of the more active parts may have to be handled
remotely, while other components will be at activity
levels where time, shielding, and distance can be
utilized. The accelerator beam tube at the Los Alamos
Meson Physics Facility (LAMPF) is 6 ft above the
floor and, therefore, a worker's head would probably
receive the greatest exposure. A person working
along the beam tube and associated apparatus will, on
the average, have his head 50 cm from the induced
activities, while his chest is at 60 cm. It then seems
reasonable to measure the dose to the head.

EYEGLASS MEASUREMENTS USING TLD's

Small area radiation exposure measurements
have been substantially implemented by the development

of thermoluminescent dosimetry. In this particular application, miniature dosimeters have been placed in various positions in eyeglass frames to measure gamma and neutron exposure to the head. These mini TLD's are commercially available and have 10 mg of LiF phosphor vacuum sealed in a glass capillary 1.4 mm in diameter and 12 mm long.

To discriminate between gamma and neutron dose, advantage is taken of the high thermal-neutron cross section of ^{6}Li (950 barns) and low thermal-neutron cross section of ^{7}Li (0.036 barns). For this purpose, both kinds of TLD's are mounted on the glasses. Since the gamma-sensitivity of these materials is about equal and they are simultaneously exposed in the same geometry, any gamma component of dose that might accompany the neutrons is read on the ^{7}Li and subtracted from the ^{6}Li to obtain the neutron dose.

For fast neutron dosimetry, a moderator is used as the ^{6}Li is primarily a thermal-neutron detector. In this case, a plastic man was used to thermalize the neutrons. This plastic man is fabricated as a molded plastic coat surrounding a real human skeleton such that certain sections can be used separately or, if desired, as a complete man. He weighs 136 pounds when filled with water, which is somewhat less than the 154-pound standard man. The head section (Fig. 1), which includes part of the chest, weighs 6.75 pounds when empty and 20.75 pounds when filled with water. It is this section that was used for the eyeglass study. Two mini-TLD dosimeters, one ^{6}Li, and one ^{7}Li, were placed at each of several positions in the eyeglass frames (Fig. 2). The positions were at the bridge of the nose, at the bottom of the eyeglass

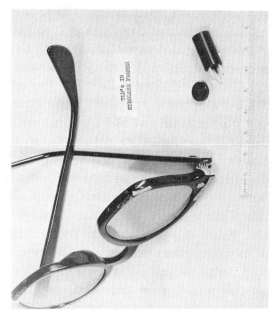

Figure 2

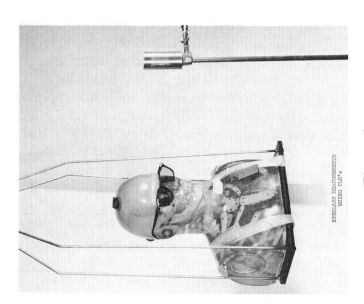

Figure 1

frame next to the nose, at the eyeglass temple bar junction, and next to the temple on the temple bar. In addition, a packet of six (three of each type) was taped on the eyeball. An entrance port allows access to the center of the skull. Another package of six was placed in this position as a reference base for all head exposures. Also, another package of six was placed in a film badge on the man's chest.

The man was positioned on a rotating mechanism of two revolutions per minute. Neutron sources of various energies were then positioned at one meter from the man at head height for a 6-ft man. A conversion to rem dose is based on cross calibration to the Hankins 10- in. sphere, Model 6, neutron monitor.[1] This instrument was positioned at one meter from the source and the mrem/h noted. Then the plastic man's head was likewise positioned and integrated doses were obtained by timed exposures.

Figure 3 shows the results of many exposures, varying from 30 to 600 mrem. As would be expected, the center of the skull showed the most consistent results and, because of the greater moderation, the greatest sensitivity. The eyeball and the temple are very much alike and between five and six times less sensitive than the center of the skull. The bridge of the nose, the eyeglass temple bar junction, and the bottom of the eyeglass frame gave erratic results. This was attributed to our inability to locate the glasses in the same position with respect to head cavities for each set of measurements. The eyeball and the temple were pretty much the same for all tests. The chest was the least sensitive, even though corrections were made for the greater distance from the sources. No doubt, the lung cavities contributed to this by absence of moderating material.

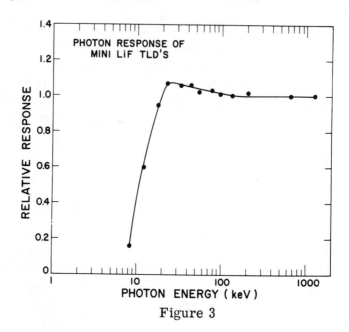

Figure 3

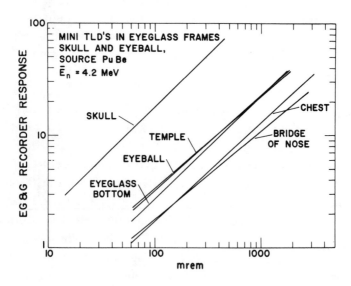

Figure 4

The plastic man was again fitted with eyeglasses, with dosimeters at the temple bar and eyeball, and exposed to ^{137}Cs gamma rays in the same manner that they were exposed to neutrons. Dose rates were determined with a Nuclear-Chicago Cutie Pie and timed exposures were again used for integrated doses. There was no demonstrable difference between the temple bar and the eyeball, and, as expected, there was no appreciable difference in the ^{6}Li and ^{7}Li responses. The energy dependence of these mini TLD's, shown in Fig. 4, is very good from 20 keV to 1.2 MeV.[2]

These measurements have demonstrated a method of rapid and simple assessment of personnel exposure to gamma radiation. The neutron sensitivity is too low for applying this method to personnel neutron monitoring.

APPLICATION TO LAMPF
HEALTH PHYSICS PERSONNEL DOSIMETRY

All personnel working on radioactive accelerator components will be required to wear glasses. A small packet of three TLD's will be attached to the temple bar. These dosimeters will be read at the end of the work period and documented on the worker's radiation history card maintained by Health Physics at the work area.

For consistent results in any TLD measurements, extreme care must be used so that all readings are obtained under the same condition. The annealing process is important in that all dosimeters are treated alike. New TLD's are heated at 400°C for one hour.

Thereafter, prior to use, all mini TLD's are annealed as follows:

   a. Heated to 400°C for 15 minutes.

   b. Exposed simultaneously to 200 mR ^{60}Co.

   c. Heated for 2 min at 135 C.

   d. Peak height of the glow curve is read out and dosimeters assigned correction factors in increments of 5%. See Fig. 5.

   e. For all measurements, dosimeters with the same correction factors are used.

AREA MONITORING WITH TLD's

Another application for these miniature TLD's is in area monitoring of the mixed radiation fields around pulsed accelerators. For this purpose the TLD's are placed in a polyethylene moderator. They are of particular use in accelerator dosimetry where radiation rate effects and instrument perturbing radio-frequency fields are of concern.

Tentatively, we have selected a 10" polyethylene sphere as the neutron moderator. Subsequent determination of the neutron spectra outside the biological shield may dictate a different diameter sphere. In the center of the sphere in a polyethylene wand six ^{6}Li and two ^{7}Li mini TLD's are positioned (Fig. 6).

Figure 7 shows the response to two neutron sources. In this manner, we have measured neutron exposures from 6×10^{5} n/cm^{2} to 10^{8} n/cm^{2}, with measurements valid to $\pm$ 30%.

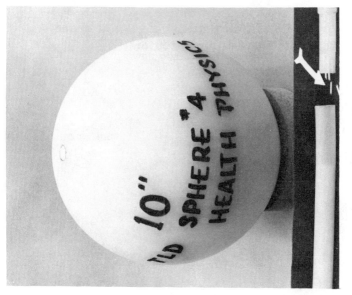

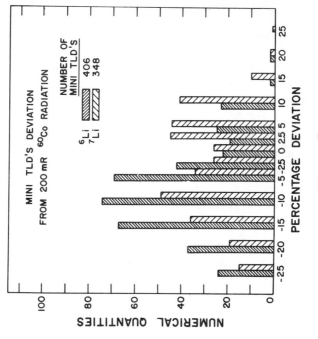

Figure 5

Figure 6

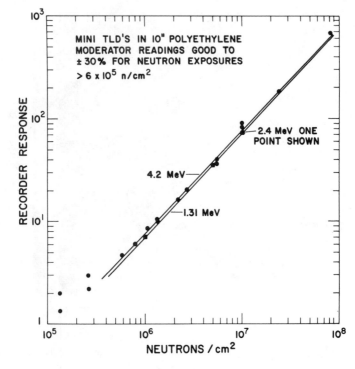

Figure 7

The fading characteristics of LiF dosimeters is 5% per 12 weeks at room temperature.[3] This will enable us to document radiation exposures in work areas for any time up to a quarter of the year. This technique is not intended to replace film badges but rather to complement them. A comparison of the two dosimetric systems will be made.

NEUTRON SPECTROMETER USING TLD's

Preliminary investigations suggest that by using various diameter polyethylene spheres and mini

TLD's as dosimeters it may be possible to measure the average neutron energy. Polyethylene spheres from 2" to 18" in diameter are being used. By plotting the sphere diameter vs relative response, Fig. 8, it is noted that the curves peak at different sphere diameters, the diameter being a function of the neutron energy. For $\bar{E}_n$ = 0.5 MeV the peak occurs in the 6" sphere, for $\bar{E}_n$ = 2.4 MeV the peak occurs in the 7" sphere, for $\bar{E}_n$ = 4.2 MeV in the 8" sphere, and so on. The data is not complete for the 8.0 MeV and 20.0 MeV neutron sources. Additional measurements at other energies are being made.

SHIELDING MEASUREMENTS WITH TLD's

As a prototype for the Meson Physics Facility, an electron accelerator has been built that accelerates electrons to 25 MeV with an average current of 1 mA. The shielding for this accelerator is built out of steel counterweights acquired from abandoned missile sites. The shield wall is 39" of steel and 12" of concrete.

In stacking these counterweights, cracks in the shield wall were unavoidable but fortuitous for many different kinds of measurements by Health Physics. In one of these cracks, perpendicular to the beamline and downstream over two meters from the target, a 12' wand of lucite was inserted at beam-tube height. Two mini TLD's (one each ^{6}Li and ^{7}Li) were positioned every four inches in this wand. The gamma and neutron exposures of these TLD's, shown in Fig. 9, were obtained by running the machine for 20 minutes. In the beam channel, the gamma radiation ranged from

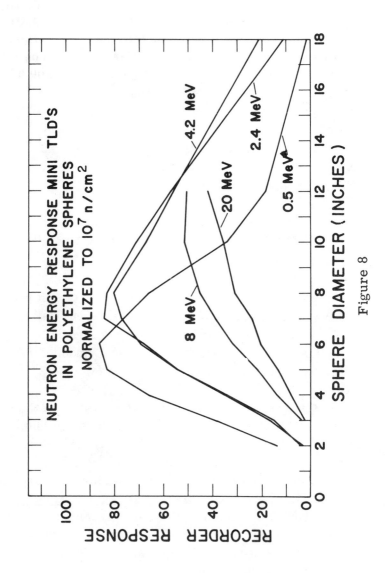

Figure 8

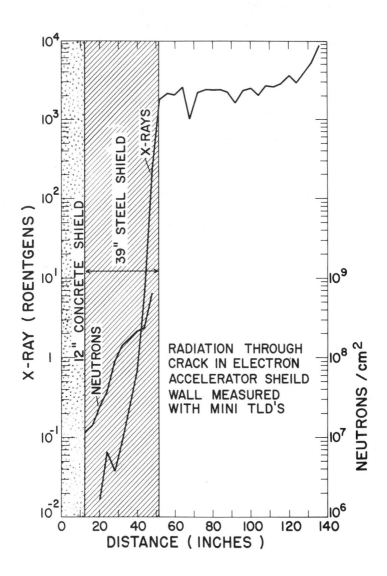

Figure 9

9000 R on beamline to 1800 R at the inside face of the shield. The gamma radiation attenuation through the iron covers 5 decades. In the beam channel, because of the high gamma-to-neutron ratio, small differences between large numbers precluded neutron assessment by using TLD's. However, in the steel shield wall the neutrons could be measured and were arbitrarily interpreted as thermal neutrons which ranged from 5.9×10^8 to 1.25×10^7 n/cm^2. No measurements were available in the concrete as the last four pairs of dosimeters were broken, due to some external stress on the wand. The area outside the concrete measured with a Model 6 LASL neutron survey instrument indicated neutron levels of 1 mrem/h with no detectable gammas (Victoreen 440-RF) while this run was in progress.

CONCLUSION

TLD's are very useful for radiation measurements around an accelerator. For neutron dosimetry, attention must be given to scattering and moderation by surrounding media. Use in mixed radiation fields should be limited to areas where the gamma-to-neutron ratio is small.

ACKNOWLEDGMENTS

The author acknowledges the cooperation and assistance of Bruce Riebe and Harry Craig.

REFERENCES

1. Dale E. Hankins, "The Multisphere Monitoring Technique," La 3700.

2. Bruce J. Krohn, William B. Chambers, and Ellery Storm, "The Photon Energy Response of Several Commerical Ionization Chambers, Geiger Counters, and Thermoluminescent Detectors," Los Alamos Scientific Laboratory report, in preparation.

3. N. Suntharalingam et al, Phys. Med. and Biol., Volume 13, p. 97 (1968).

WALK-OVER MONITOR FOR ALPHA AND BETA-GAMMA CONTAMINATION*

William F. Splichal, Jr.
Savannah River Laboratory
E. I. du Pont de Nemours and Co.
Aiken, South Carolina

A walk-over hand and shoe monitor was developed to quickly screen personnel for alpha and beta-gamma contamination when leaving a regulated work area. A walk-over bridge monitor was selected because it is easy to use and therefore is readily accepted by personnel, and it also allows a large number of employees to be monitored quite rapidly which is particularly important at shift changes. Other walk-over or walk-through personal contamination monitors have been reported, but none monitored for alpha, as well as beta-gamma contamination.

---

*The information contained in this article was developed during the course of work under Contract AT (07-2) -1 with the U.S. Atomic Energy Commission.

911

The monitor (Fig. 1) allows 60 people per minute to be checked, which is considerably faster than conventional stand-on monitors. Approximately 4×10^3 dis/min of alpha, 1.6×10^4 dis/min of ^{204}Tl (0.77 MeV beta), or 2.2×10^6 dis/min of ^{54}Mn (0.84 MeV gamma) on hands or shoes of a person walking across the monitor at a normal pace triggers an alarm.

DETECTORS

The detectors selected for the monitor are simple and inexpensive due to the large number required. However, they provide a high degree of reliability and detection sensitivity.

Beta-Gamma Detector

Scintillation detectors and GM tubes were considered for beta-gamma contamination detectors. The less expensive 900-volt halogen-quenched GM tubes were selected. They are small in size but can be as sensitive to beta-gamma radiation as conventional type scintillation detectors if more are used for a given area. Four of these tubes (Fig. 2) are evenly spaced in each of the twelve beta-gamma shoe detector sections, and in each of the four beta-gamma hand detector sections in the hand rails. Each GM tube was placed in a 1/4-inch-thick channel-shaped lead shield to reduce background and to permit only radiation from above to reach the detector. The 64 small nonselected GM tubes have more than a 100-volt operating plateau which eliminates the necessity of individual voltage settings. Each tube is individually decoupled with a

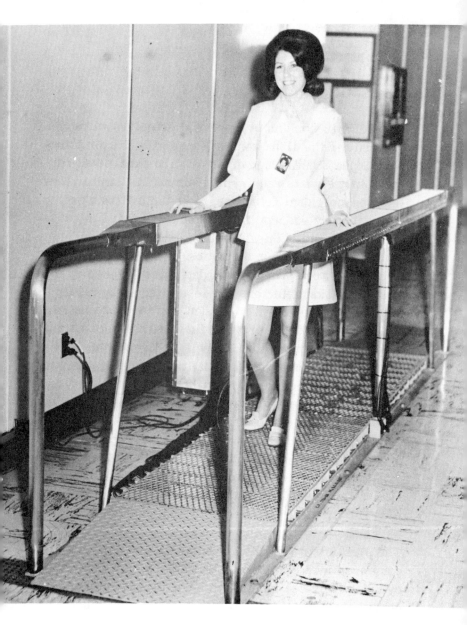

Fig. 1. Person Walking Over Bridge

simple RC network to prevent cross-talk which could give false pulses at the input of other detector sections.

Alpha Detector

Scintillation and proportional detectors were considered for the alpha contamination detectors. Large area alpha scintillation detectors are expensive, but the output signals require less amplification than those from proportional detectors. The scintillation detectors are very susceptible to light leaks that can cause operating failure; this was considered an inevitable problem with the shoe detectors.

Large area proportional detectors when compared to scintillation detectors are less affected by small pin holes in the detection window and less expensive for large area detectors, but they require more signal amplification and are sensitive to microphonics and humidity. Using a proportional type counting gas was out of the question because of the high cost of continually purging the large area detectors and the safety aspects of using flammable counting gas. Recognizing these factors, proportional detectors using air, which on first thought might be considered one of the worst possible choices of counter gas (especially in the humid climate of South Carolina) were selected.

The inherent disadvantages and problems of using air as the counter gas were overcome or minimized. The erratic behavior due to humidity normally encountered with air proportional detectors was minimized by using oversize insulators, wide spacing between center electrode wires, polished interior conducting surfaces, and a lower operating potential of 2400 volts instead of the normal 3800 volts. Microphonic effects were reduced to a point where only vigorous stomping on the shoe section causes false alarms by prestretching the "Mylar"* windows and using an input rise time constant much shorter than the rise time of the pulses due to microphonics. Small pin holes caused by grit and other small foreign material do not affect the detectors and can be patched with small pieces of tape.

A series of 12 proportional detectors (Fig. 3), each 4 x 24 x 5/8-inch, provide a total area of 8 sq ft to detect alpha contamination on shoes. The four hand alpha detectors, each 5 x 24 x 5/8-inch, are in the two hand rails. The air proportional counters (Fig. 4) have 0.001-inch-thick tungsten center electrode wires supported around "Teflon"* insulators. Each alpha detector is covered with a 0.00015-inch-thick doubly aluminized "Mylar" window on a removable frame which can be quickly changed if damaged or contaminated.

BRIDGE

The overall length of the bridge is 15 feet, which insures that both feet come in contact with each of the five-foot long alpha and beta-gamma shoe detector sections. A thin expanded metal grill of 00.125-inch-thick steel is used as the bridge floor to protect the 24 shoe detector sections. Two-foot long inclines on each end of the bridge lead up to the 3-inch-high bridge

*Trademark, E. I. du Pont de Nemours and Company.

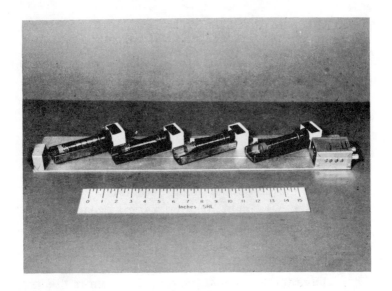

Fig. 2.   B-γ Detector Section

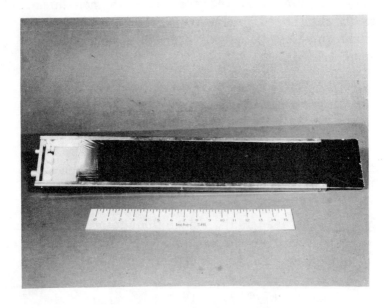

Fig. 3.   α Detector Section

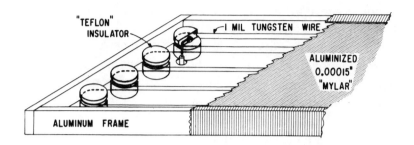

Fig. 4. Exploded View of α Detector Section

floor. Spring loaded linkage mechanisms below each shoe detector allow the detectors (Fig. 5) to be dropped for removal from the side of the bridge for servicing. Each hand rail is divided into two sections and the hinged covers (Fig. 6) over the detectors can be lifted over for servicing. Two polyethylene foam pads are placed under each hand detector to hold them at the top of the hand rails against the stainless screen which protects the hand detectors.

CIRCUITRY

A metal cabinet attached to one side of the bridge (Fig. 7) contains the transistorized circuitry, and has six indicator lamps to indicate if the contamination is alpha or beta-gamma and if the contamination is on the left hand, right hand, or shoes. When the activity level reaches any of the preset levels, a contamination alarm system is activated for 5 seconds before automatically resetting. This system includes the six contamination lamps, a high pitch audio oscillator, and a large overhead electric sign reading "Decontamination Required".

Negative pulses from the air proportional counters, (Fig. 8) which operate at 2400 volts, are amplified by a charge-sensitive amplifier. High-gain charge-sensitive amplifiers were used because of the relatively large capacitance of the proportional detectors and associated coaxial signal cables and to make up for the amplification lost in the proportional detectors when operating them at a low potential. Amplified pulses feed a monostable multivibrator that drives an integrating circuit. A count rate above the preset

Fig. 6. Hand Section Being Removed

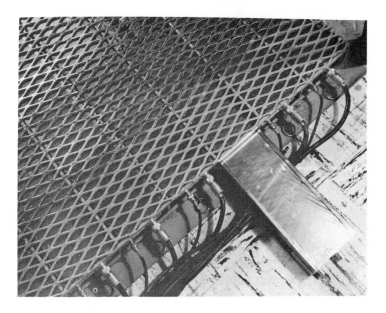

Fig. 5. Shoe Section Being Removed

Fig. 7. Photo of Electronics Showing Trouble
Shooting Lights.

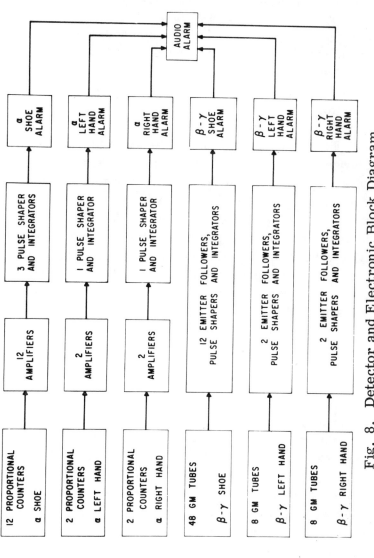

Fig. 8. Detector and Electronic Block Diagram.

alarm level is determined by the integrating circuit that triggers an alarm switching circuit to actuate the alarm. The period of alarm is adjustable and can be set to any desired length of time.

Negative pulses from the GM tubes are impedance matched by an emitter-follower to a monostable multivibrator, and then are fed through circuitry identical to that in the alpha section.

Two regulated power supplies prevent cross-talk between the analog circuitry, which consists of all pulse amplifiers, and the digital circuitry, which consists of all pulse shaping and switching circuits. The monitor is automatically powered without interruption by two 6-volt batteries in event of an AC power failure. High voltage for the GM tubes and proportional counters is provided by two blocking oscillators; one oscillator feeds a voltage doubler and the other a voltage quadrupler.

SERVICING

A unique feature of the monitor is the ability to remove, repair, or replace any of the 32 detectors or 46 printed circuit boards without interrupting monitoring. Because of the difficulty in trouble-shooting a monitor with so many detectors, a panel of 32 indicator lamps on the side of the electronics chassis simplifies the task of quickly locating a defective detector. These lamps indicate the condition of each detector section with a light flash for each ionizing event. Separate detector power switches allow individual defective detectors to be switched off until they can be repaired.

Buildup of debris and dust on the shoe sections is easily cleaned off periodically with a vacuum cleaner. A heavy duty rubberized door mat placed on the lead-up ramp of the bridge removes a large portion of foreign material on shoes before personnel walk over the bridge. Less maintenance is required by having the bridge oriented so that personnel walk over the beta-gamma shoe detectors first. After walking over the rubber mat, any remaining debris can fall into the beta-gamma shoe detector sections which are much less susceptible to foreign material than the alpha shoe detectors. The frequency of "Mylar" window replacement on the alpha detectors is approximately one window per week and this usually takes only a few minutes. In most cases, damaged windows are caused by large punctures from nonradioactive metal turnings adhering to the shoes of machine shop workers. None of the 0.001-inch tungsten center electrode wires in either the alpha hand or shoe detectors has been broken in the 10 months the monitor has been in service. The monitor is used ~2000 times each day and the average maintenance required is approximately one hour each week. The greatest portion of the maintenance is inspecting, patching, or replacing "Mylar" windows.

EVALUATION

Personnel acceptance of the monitor has been excellent. Although the detection level is not as low as other type monitors which require more effort and time to use, this monitor insures that every person who leaves a regulated area does so without significant amounts of contamination on the hands or shoes. The monitor has proved effective in detecting unsuspected cases of contamination and has been beneficial in maintaining good contamination control.

CALIBRATION AND FIELD USE OF IONIZATION* CHAMBER SURVEY INSTRUMENTS

W. P. Howell

and

R. L. Kathren

Battelle Memorial Institute
Pacific Northwest Laboratory
Richland, Washington

ABSTRACT

Measurement accuracy of ion chamber survey instruments in field survey situations is affected by both the calibration procedures and the geometric relationship between the radiation source and the chamber. This paper describes the fundamentals of the calibration

---

*This paper is based on work performed under United States Atomic Energy Commission Contract AT(45-1)-1830.

method used at Hanford for ion chamber instruments, and the resulting capabilities of the instruments in field survey situations. Additional corrections to calibrated instrument results are ordinarily required when measuring small beams, radiation fields associated with large beta sources, and radiation levels where the chamber is in contact with the source. The method of determining the required corrections, and the values of the correction factors are described for one widely-used CP ion chamber survey instrument.

INTRODUCTION

Ionization chambers are widely used as detectors in protable radiation monitoring instruments. The advantages of the ion chamber are many, and include wide range, the ability to utilize simple circuitry, and low voltages, and ease of chamber construction, permitting numerous geometries, and the use of air or tissue equivalent walls for minimal energy dependence.

Many ion chamber instruments are affectionately known as "cutie pies", and this phrase has joined other common biological entities-pig and rabbit-in the lexicon of the nuclear world.* The term cutie pie (CP) was orginally used to describe ion chamber instruments with a characteristic pistol shape. However

---

*The etymology of the nuclear term cutie pie is obscure. However, the term appears to have originated from a combination of need for security with respect to the use of the instrument and the mathematical symbols Q, t, and π in which Q was charge, t, time, and π, the geometry factor.

in recent years, the term has been applied to virtually any portable survey meter utilizing a non-pressurized ionization chamber.

This paper will describe the application of a specific cutie pie instrument to field monitoring situations at the Hanford site. After a brief description of the instrument, the actual procedures used in the field will be presented, with particular emphasis given to geometry corrections and measurement of dose rates from mixed beta-photon fields. Finally, the paper will conclude with a discussion of some of the limitations of the Hanford-type cutie pie. While much of the information presented here is specific to a single instrument, an appreciable portion, including the basic concepts, should be applicable to other instruments utilizing similar unpressurized ion chambers.

THE HANFORD CP

The Hanford "Cutie Pie" (Fig. 1) is a pistol shaped portable dose rate meter using an ion chamber as a detector. The ion chamber is constructed from a piece of phenolic plastic tubing, 3 in. in diameter and 5.7 in. long, providing an effective chamber volume of 40.2 in.^{3} (658 cm^3). The walls of the chamber are 0.125 in. (440 mg/cm^2) thick. As shown in Fig. 2, one end of the tubular chamber butts against the body of the instrument, while the other is fitted with a disc of Mylar or rubber hydrochloride 7 mg/cm^2 thick. A window shield, 440 mg/cm^2 thick, is also provided. All interior surfaces of the chamber are coated with a conductive graphite surface.

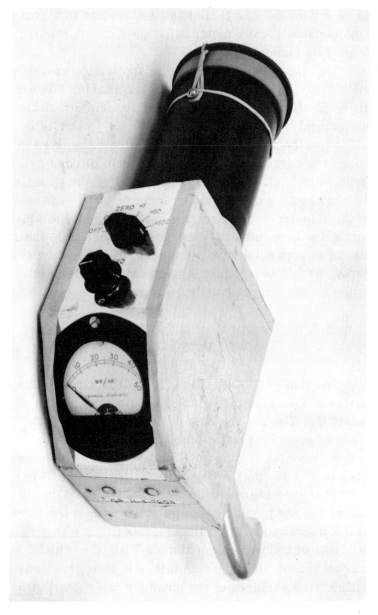

Figure 1

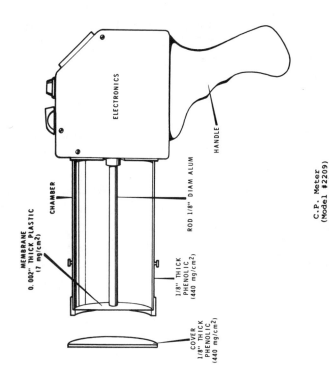

C.P. Meter
(Model #2209)

Figure 2

A center electrode, 0.125 in. in diameter, runs
the full length of the chamber and is utilized for ion
collection. With the 658 cm^3 chamber, a dose rate of
1 mR/hr provides 1.4 x 10^9 ion pairs or a current of
about 3 x 10^{-14} amperes. This is about the minimum
level of detection available with a simple electrometer-
type circuit. Hence, the instrument has been provided
with ranges of 0-50, 0-500, and 0-5,000 mR/hr.

The energy dependence of the Hanford CP in measurement of electromagnetic radiation has been
determined by comparison with a free air ionization
chamber and NBS calibrated R-meters. Figure 3
shows the results of the study. Between 150 keV$_{Eff}$
and 3 MeV$_{Eff}$, response is essentially flat, irrespective
of chamber orientation. Below 150 keV$_{Eff}$, the response rises, peaking at 100 keV$_{Eff}$. At this point, the
closed window response is about 40-50% high, relative
to 1 MeV. Below this point, response falls, so that
at 17 keV$_{Eff}$, the response is about 10% low. While
this error is significant, it can be readily corrected,
permitting the instrument to be utilized for monitoring
the dose rate from bare ^{239}Pu, an important capability
at many sites, including Hanford. So important is
this particular capability, that it can override numerous other desirable features, including a flatter energy
response above 40 keV$_{Eff}$.

Because of the 440 mg/cm^2 wall thickness, beta
particles with energies below 1.1 MeV will be excluded
from the chamber with the window closed. However,
with the window open, the CP will monitor all the betas
that would penetrate the dead layer of the skin. Therefore, if suitable corrections are made for geometry,
the CP can be used for beta dose rate determinations
in the field, as discussed below.

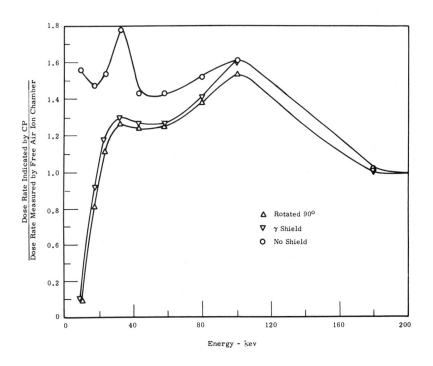

Figure 3

Hanford CP instruments are typically calibrated with a standardized source of ^{226}Ra in equilibrium with daughters. [1] The instrument is checked at three points on all ranges and calibrated to +10, -5% of the true exposure rate. A well calibration system is used, along with a standard positioning jig to ensure reproducible geometry. In addition to the actual calibration, the instrument is driven off scale on all ranges to ensure that it will not fail to respond in a high radiation field.

The importance of a reproducible calibration geometry cannot be too heavily stressed. Many erroneous instrument readings can be directly traced to faulty or inappropriate calibration techniques. However, calibration is not the only factor that affects the use of a dose rate instrument, and the more common serious factors will be considered from the standpoint of field measurements.

FIELD USE OF THE CP

In field survey situations, the four factors which most often introduce errors into ion chamber instrument measurements are chamber energy dependence, variations in source size and instrument to source distance, radiation beams which do not irradiate all of the chamber gas, and chamber limitations on the angular acceptance of beta radiation.

Using the energy dependence curve illustrated in Fig. 3 above, meter indications in situations involving gamma or X-ray energies below 150 KeV can be corrected when necessary, if the effective energy of the radiation is known, and if it is monoenergetic.

In unknown situations, or those involving a mixture of high and low energies, more sophisticated instruments and methods must be used for dose rate evaluations. For example, very high energy photons might be encountered at an accelerator or reactor, with the predominant photon energies greater than 3 MeV. Under these circumstances, the CP will under-respond, by as much as 50% at commonly encountered energies. In these cases, a method of measuring kerma should be utilized.

All of the other error-causing factors involve some degree of non-uniform ionization of the chamber gas, and meter indications lower than the true dose rate will result. Since the CP is calibrated using radium gammas under conditions which provide a uniform degree of ionization throughout the chamber, the instrument will indicate true dose rates only under survey conditions of a similar nature. Thus, in the case of measurements near the surface of relatively small sources, erroneous readings will occur. Where the error is small, the situation is of little concern, but errors can be as large as a factor of 100 where the situation involves a contact measurement on a small beta source.

The response of the CP instrument under conditions encountered in field monitoring situations has been defined in four separate studies by Hanford personnel. [2,3,4,5] The first of these was designed to determine true contact dose rates on cylindrical sources, and to provide suitable corrections to instrument readings for these measurements.

The second study was designed to define the angular response of the CP in measurement of extended beta radiation sources. The study covered source-to-instrument distances between 4 and 36 inches for source diameters between 6 and 72 inches. Correction factors were obtained from the ratio of the total integrated dose rate obtained with the forward axis of the chamber intersecting the center of an incremental source to the total integrated dose rate obtained with the forward axis of the chamber intersecting the center of the integrated source.

The third study was designed to define the actual response and error in measurements of plane circular gamma radiation sources of various radii. Measurements of an incremental source and numerical integration were used to derive true contact and near contact dose rates. Film dosimeter data supplemented the mathematical calculations. Correction factors were derived for sources having diameters between 0.2 and 144 inches.

The final study was designed to develop a family of correction factor curves for both beta and gamma radiation for all monitoring conditions, utilizing the results of the earlier studies and additional measurements. A small table of the most useful numerical values obtained from the correction factor curves was assembled. Copies were provided to field personnel for attachment to the CP instrument case. Figures 4-6 are the results of these studies.

In field use, the correction factors apply to commonly encountered cylindrical and plane circular sources, radiation beams, and extended beta sources. Since the source diameter may not be readily apparent in a field monitoring situation, the surveyor is often required to scan the area and estimate the source diameter. It should be noted that ambient beta fields, as might be encountered in a room with walls, floors,

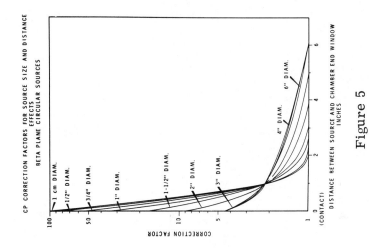

Figure 5

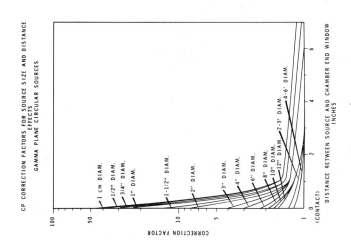

Figure 4

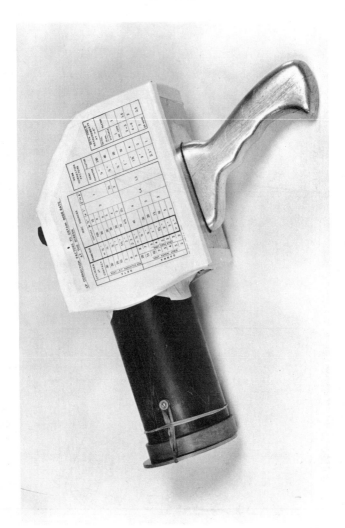

Figure 6

and ceilings contaminated with a beta emitter can be measured with the CP, utilizing beta field corrections. However, these situations must be evaluated carefully, since the distance from source to detector is variable, and some of the more energetic betas may penetrate the side walls of the chamber. Where the latter occurs, the beta field correction cannot be used.

In a typical field situation, Pacific Northwest Laboratory personnel normally approach an unknown survey situation with the CP window shield open. The following procedure is used in determining the correct dose rate:

1. Determine meter deflections, shield open and shield closed.
2. Determine source size.
3. Apply the appropriate correction factor(s) to to the meter deflections, according to the table in Fig. 6.
4. The corrected closed shield measurement is the penetrating dose rate (normally only electromagnetic radiation).
5. The difference between the shield open and shield closed meter deflections, when corrected is the non-penetrating dose rate.
6. The sum of the measurements in steps 4 and 5 is the total skin dose rate, including both penetrating and non-penetrating radiation.

This procedure is relatively simple, and has been used successfully by monitoring personnel at Hanford for about ten years.

The Hanford CP, as is true of other instruments of its type, has numerous limitations. However, if intelligently used, the instrument can provide the operational health physicist with reasonably accurate dose rate estimates under field conditions.

REFERENCES

1. R. L. Kathren, and H. V. Larson, "Radiological Calibration and Standardization for Health Physics," BNWL-SA-1600 (1967); Health Physics, 15 (in press).

2. G. L. Helgeson, "Surface Dosimetry and Effective Energy Calculations," HW-41439 (1956).

3. I. C. Nelson, "Beta Field Dose Rate Determination With the CP Dose Rate Meter," HW-40809 (1956).

4. J. F. Evans, "Surface Dose Rate Measurements," HW-57339 (1958).

5. L. A. Carter, unpublished data.

PERSONAL MONITORING AT A FOUR MEV VAN DE GRAAFF ACCELERATOR WITH TISSUE-EQUIVALENT POCKET DOSIMETERS

D. J. Sreniawski and N. A. Frigerio
Argonne National Laboratory*
Argonne, Illinois

ABSTRACT

The operators and users of a four MeV de Graaff accelerator historically had been assigned film badges using Kodak NTA film as personal monitors for neutron radiation. When a review of their cumulative neutron exposures as reported by film badges was not supported by radiation survey records of the work areas, a study was made to determine the reason. The study found that the neutron average energy was less than 200 keV in areas occupied by the accelerator personnel. Neutrons of this energy are below the region of effective interpretation for the Argonne National Laboratory film badge. A search was initiated for a supplement to the film badge resulting in the adoption of a tissue-equivalent ionization chamber pocket dosimeter. This

*Work performed under the auspices of the United States Atomic Energy Commission

paper describes the tests made to determine its energy dependence.

INTRODUCTION

The operators and users of the four MeV Van de Graaff accelerator of the Argonne National Laboratory (ANL) Physics Division had been assigned film badges containing Kodak nuclear track film type A (Kodak NTA) for neutron personal monitoring. Each individual's badge was processed every four weeks and the results recorded. The lower limit of dose which is reported is 50 mRem on a PuBe spectrum calibration.[1] A review of the film badge records showed no accelerator operator had received a reportable neutron exposure (50 mRem in any four-week film period) for at least the last three years. At the same time measurements of the stray radiation field during some accelerator experiments indicated neutron dose equivalents up to 150 mRem for a four-week film period. This dose equivalent was measured at the accelerator control panel using a tissue-equivalent ionization chamber and converting from the absorbed dose (rads) by multiplying by a quality factor (QF) of 8.[2] A determination of the stray neutron average energy showed why the film badges failed to detect neutron irradiations. The neutron average energy in these areas was less than 200 keV, well below the sensitivity of the Kodak NTA film. This film requires a neutron energy of 450 keV to produce a minimum recoil proton track length of four grains in the emulsion.[3]

FAST NEUTRON AND GAMMA POCKET
DOSIMETER SYSTEM

A search was made for a supplement to the film badge that would be sensitive to fast neutrons below 200 keV. A commercially manufactured fast neutron and gamma pocket dosimeter system was eventually chosen. The system consists of a tissue-equivalent dosimeter sensitive to fast neutrons and gamma plus a neutron-insensitive dosimeter. Their general construction is similar to conventional pocket dosimeters with one distinction (Figure 1). [4] The walls of the ionization chamber of the first unit are made of tissue-equivalent plastic such that the response is based on the effect of collisions between fast neutrons and atoms of soft tissue of the composition $(C_5H_{40}O_{18}N)_n$. The

THE DOSIMETER

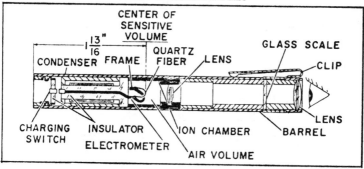

Figure 1.

ionization chamber and surrounding insulation of the neutron-insensitive dosimeter had the elemental

hydrogen reduced to less than 2% to suppress the proton recoils and n, γ reactions. Gamma response of
the neutron-insensitive unit is based on carbon to match
the tissue-equivalent dosimeter's gamma response.
The first collision dose of fast neutrons is determined
by subtracting the reading of the neutron-insensitive
dosimeter from that of the tissue-equivalent unit. Both
dosimeters have a 200 millirad full scale.

The manufacturer claimed a response of $\pm 20\%$ of
the actual dose for neutrons in the 20 keV to 15 MeV
range.

NEUTRON ENERGY DEPENDENCE TEST

To verify neutron energy dependence, the tissue-
equivalent dosimeters were irradiated with neutrons
of the $Li^7(p, n) Be^7$ reaction using a 4 MeV Van de
Graaff accelerator. The neutron energies were approximately 88 keV, 398 and 891 keV. Four groups
of eight dosimeters per group were irradiated at each
of the three neutron energies. The neutron dose
ranged from 138 to 262 mrad and was calculated by
knowing the neutron flux (measured with a fission
counter) and the neutron energy (accelerator instrumentation and output from accelerator computer program gives this information). The tissue-equivalent
dosimeter response to fast neutrons ranged from 0.69
to 0.87 when normalized to a 1 mrad dose (Figure 2).
While this response for the limited number tested (32)
did deviate somewhat from the manufacturer's specification, the deviation was not too large for our purposes.

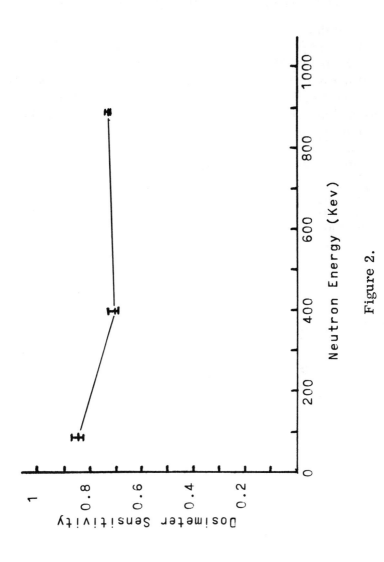

Figure 2.

SOME OTHER DOSIMETER PROPERTIES

In the course of the evaluation some other properties of the dosimeters were considered. Twenty
tissue-equivalent dosimeters made by the same vendor
but not part of the 32 tested for energy dependence,
were exposed to a 50.2 mCi ^{226}Ra source to determine
their response to gamma exposures. Each dosimeter
received two gamma exposures of 100 mrad per exposure. The average deviation was ± 8.3%. A check
on the drift in readings (the change in reading for a
unit stored in a nonactive area) was made over a two
month period. The leakage of charge produced a drift
of less than 2 mrad per month. A greater error was
encountered in the reading of the dosimeters. This
amounted to ± 2 mrad. Figure 3 shows the reason;

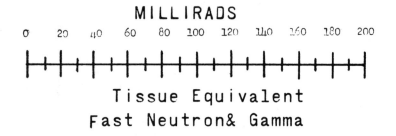

Figure 3 The Dosimeter Scale

the smallest scale division is 5 mrad. Individual interpretation of where the indicator fiber crosses the
scale introduced the 4 mrad spread. The problem is
especially acute when there is a significant gamma contribution to the mixed stray radiation. A 4 mrad error
in gamma evaluation ([QF] Quality Factor = 1) could
mask a 32 mRem neutron exposure (neutron QF = 8 :
4 mrad X QF(8) = 32 mRem). [4]

CONCLUSION

The tissue-equivalent and neutron insensitive
pocket dosimeter system, while having limitations, is
useful in specific situations. The conditions are: (1)
low gamma exposures (~1/5 of the neutron dose in
mrads); (2) a fast neutron dose of less than 200 mrad
per period of interest; (3) the neutron average energy
is known so the appropriate QF may be assigned to
permit conversion from absorbed dose (mrad) to dose
equivalent (mRem).

The principle advantage of the system is its sensitivity to a much broader neutron energy spectrum
when compared to alternate personal monitoring systems.

Adoption of the tissue-equivalent pocket dosimeter
system at the ANL Physics accelerator have not solved
all our neutron personal monitoring problems. At the
current state of the art determination of the true dose
persons receive from neutrons is still difficult. However, we consider the supplemental use of the tissue-
equivalent dosimeter system is an improvement over
the exclusive use of Kodak NTA film badges.

Table 1.

Irradiation #1, 88 keV, 150 mrad

Dosimeter #	Dosimeter Reading (mrad)	Dosimeter Reading 150 mrad
05	123	0.82
06	117	0.78
07	125	0.83
09	135	0.90
13	133	0.89
17	127	0.85
22	114	0.76
	#1A Group Average	0.83
24	127	0.85
26	129	0.86
27	136	0.91
29	123	0.82
30	140	0.93
34	139	0.93
37	123	0.82
41	130	0.87
	#1B Group Average	0.87
42	123	0.82
43	125	0.79
47	118	0.83
48	105	0.70
50	133	0.89
51	148	0.99
53	131	0.88
54	121	0.81
	#1C Group Average	0.84
55	113	0.75
59	115	0.77
62	125	0.83
65	117	0.78
66	135	0.90
70	123	0.82
71	144	0.96
73	135	0.90
	#1D Group Average	0.84
88 keV Irradiation Average		0.85

Table 2.

Irradiation #2, 398 keV, 138 mrad

Dosimeter #	Dosimeter Reading (mrad)	Dosimeter Reading 138 mrad
05	95	0.69
06	93	0.67
07	96	0.70
09	104	0.75
13	Discharge	----
17	94	0.68
22	99	0.72
	#2A Group Average	0.70
24	100	0.72
26	99	0.72
27	105	0.76
29	100	0.72
30	108	0.78
34	103	0.75
37	92	0.67
41	100	0.72
	#2B Group Average	0.73
42	104	0.75
43	92	0.67
47	106	0.77
48	81	0.59
50	104	0.75
51	106	0.77
53	96	0.70
54	97	0.70
	#2C Group Average	0.71
55	92	0.67
59	108	0.78
62	94	0.68
65	98	0.71
66	95	0.69
70	84	0.61
71	97	0.70
73	98	0.71
	#2D Group Average	0.69

398 keV Average   0.71

Table 3.

Irradiation #3, 891 keV. Group A - 262 mrad.

Groups B, C and D - 208 mrad

Dosimeter #	Dosimeter Reading (mrad)	Dosimeter Reading 262 mrad
05	197	0.75
06	187	0.71
07	184	0.70
09	204	0.78
13	Discharge	----
17	198	0.75
22	173	0.66
	#3A Group Average	0.73
		208 mrad
24	158	0.76
26	151	0.73
27	148	0.71
29	149	0.72
30	164	0.79
34	163	0.78
37	149	0.72
41	152	0.73
	#3B Group Average	0.74
42	153	0.74
43	147	0.71
47	164	0.79
48	148	0.71
50	156	0.75
51	158	0.76
53	159	0.76
54	147	0.71
	#3C Group Average	0.74
55	140	0.67
59	136	0.65
62	152	0.73
65	149	0.72
66	168	0.81
70	146	0.70
71	162	0.78
73	160	0.77
	#3D Group Average	0.73

891 keV Average   0.74

REFERENCES

1. T. A. Steele, "Use of Personal Monitoring Neutron Film Near the 50 MeV Injector of the ZGS", CONF-651109, Accelerator Radiation Dosimetry and Experience (1965).

2. Protection Against Neutron Radiation Up to 300 Million Electron Volts, National Bureau of Standards Handbook 63, p 15, 1967.

3. R. L. Lehman, "Energy Response and Physical Properties of NTA Personnel Neutron Dosimeter Nuclear Track Film", AEC Report UCRL-9513, 1961.

4. Nuclear Associates, Inc., Product Bulletin NA-18, Westbury, New York.

A LIVE TIME CONTROLLED SCANNING RATE FOR A WHOLE BODY COUNTER[a]

M. E. Sveum and P. E. Bramson
Battelle Memorial Institute
Pacific Northwest Laboratory
Richland, Washington

ABSTRACT

Scanning whole body counters tend to sacrifice sensitivity and introduce measurement error through the use of constant drive rate mechanisms which do not readily compensate for excessive electronic dead time or subject height. An electronic drive rate control has been developed at Battelle-Northwest Laboratory which largely eliminates these problems. The control is easily programmed for subject height and count duration and automatically adjust the relative drive rate to compensate for multichannel analyzer dead time. Digital circuitry, utilizing integrated

[a] This paper is based on work performed under United States Atomic Energy Commission Contract AT(45-1)-1830.

circuits, controls the speed of a heavy duty stepping motor by controlling the motor stepping frequency.

A scanning whole body counter utilizing the typical constant drive rate tends to underestimate the body or organ burden of a subject whose deposition caused abnormally high analyzer dead time. A counter utilizing the live-time control will spend proportionally more total scanning time over an area of high deposition, resulting in a desired amount of "scanning live time" for which the counter calibrations apply.

Internal Dose Evaluation of Battelle-Northwest operates several whole body counters of the "shadow shield" or scanning design as shown in Fig. 1.[1] There are also a number of whole body counters throughout the U. S. and Europe which utilize variations of the scanning geometry.[2] This type of counter usually employs a constant speed drive to control the scanning rate and the scanning period is normally adjusted to coincide with a multichannel analyzer counting duration of a few minutes.

In routine "low level" whole body counting the constant speed method is adequate and the completion of the analyzer count is easily coincident with the scan completion. A problem arises when relatively large radionuclide depositions, say microcurie amounts, are encountered in whole body counting; particularly if the deposition is localized. In this situation there may be a significant increase in analyzer dead time where dead time is the time the analyzer is processing pulse data during which it cannot accept additional pulses. Since the source is moving past the scintillator, this dead time increase can reduce the magnitude of data stored by the analyzer. In other words, as dead time increases, the count rate per unit activity

Figure 1.

decreases and since the source is moving past the detector, the total number of counts per total activity is
reduced. As a result, the radionuclide deposition in
a subject may be underestimated by a large factor
depending upon how large the dead time percentage
becomes.

Ideally, an increase in analyzer dead time due to
increased counting rates would cause a corresponding
decrease in sled travel rate, that is, a 5 percent increase in dead time would decrease the the sled travel
rate by 5 percent. The decrease in sled travel rate
assures that the detector spends the proper amount of
"live analyzer time" in the vicinity of radionuclide
deposition. Since the "ideal" sled drive rate is proportional to the live time of the counting system, a
logical source of drive-rate timing is the analyzer
live timer. Many multichannel analyzers produce a
timing pulse train whose frequency is proportional to
the analyzer live time, the highest frequency corresponding to clock time or, in other words, 100 percent
live time. As dead time increases the frequency of
the timing pulse train decreases in proportion.

Figure 2 presents a block diagram of a circuit
which utilizes the live-time pulse train from a multichannel analyzer, in this case an RIDL 400 channel
model, to control the stepping frequency of a heavy
duty stepping motor. The stepping motor was chosen
because of its inherent ability to turn and hold a fixed
amount upon receiving a power pulse. The number of
power pulses delivered to the motor controls the distance scanned and the frequency of the power pulses
determines the rate of scan. A one megacycle pulse
train is generated by a crystal oscillator and pulse
shaper. This pulse train is gated by the analyzer

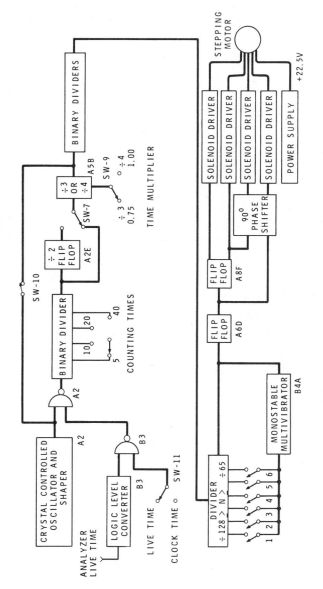

Figure 2.

live-time pulse train through a logic level converter. The gated pulse train passes through a series of variable binary dividers which allow adjustment of maximum sled travel rate and total distance traveled in the selected live time period. Finally, the pulses activate solenoid drivers to power the stepping motor. Figure 3 presents a photograph of the control and stepping motor drive. Integrated circuits were used extensively throughout the control and positive action toggle switches provide for basic travel rate adjustment.

The usefulness of such a scanning drive system is apparent from Fig. 4. The solid curve presents the integrated count attained by a conventional, non-live time controlled, scanning whole body counter for a point source located in the center of the scanned region. The dashed curve illustrates the relative difference in integrated count attained for the same source and location when scanned with a live-time controlled counter. The final integrated count value, in the latter case, yields the same count per unit activity as a relatively small source which negligibly increases analyzer dead time. The accurate value of the source activity is readily calculated from the live-time controlled scan whereas an underestimate would be obtained from the ordinary scan.

The control provides for 8 scanning times from 3.75 minutes to 40 minutes. Length of sled travel is variable in small increments up to a maximum of 87 inches in either direction. A fast-travel switch is available for rapid position of the sled at the end of a short run. The stepped travel has not proven objectionable although one can detect the steps at low scanning rates; more than 10 minutes. The stepping motor becomes quite warm during operation due to

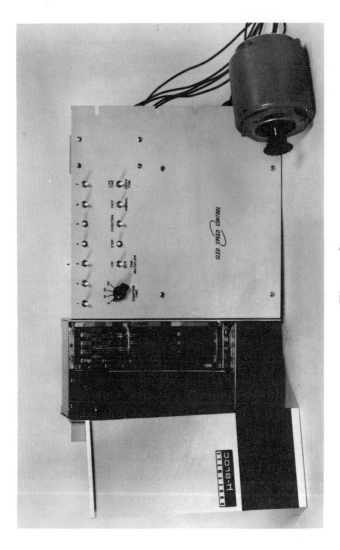

Figure 3.

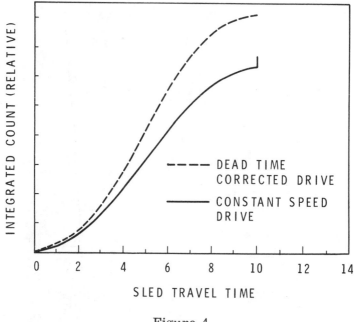

Figure 4.

the heating of the windings during the "hold" cycle but there has been no indication of motor overloading.

REFERENCES

1. D. N. Brady and F. Swanberg, Jr., "The Hanford Mobile Whole Body Counter", Health Physics 11, 1221 (1965).

2. H. G. Mehl and J. Rundo, "Preliminary Results of a World Survey of Whole Body Monitors", Health Physics 9, 607 (1963).

AUTOMATIC CONVEYOR TYPE
LAUNDRY ALPHA-BETA-GAMMA
CONTAMINATION MONITOR*

E. M. Sheen and G. D. Crouch
Pacific Northwest Laboratory
Battelle Memorial Institute
Richland, Washington

ABSTRACT

An automatic radiation monitoring system was
developed by the Pacific Northwest Laboratory for the
Federal Support Services of International Telephone
and Telegraph to scan laundered protective clothing
and automatically reject contaminated garments. This
system employs two stainless steel wire belt conveyors
that transport the wearing apparel at a belt speed of
160 in./min between two large alpha-beta-gamma
scintillation detectors that simultaneously monitor

*This paper is based on work performed under
United States Atomic Energy Commission Contract
AT(45-1)-1830.

upper and lower surfaces of each item of apparel for alpha-beta-gamma contamination. The instrument reliably sorts items for further decontamination if they are contaminated to a level greater than the equivalent of 25,000 dis/min (11 nCi) of Ra DEF or 1000 dis/min (0.45 nCi) of 5 MeV alpha emitters distributed over 100 cm^2.

INTRODUCTION

Protective clothing is worn by workers at Hanford in radiation zones to reduce the possibility of personal contamination and to inhibit the spread of radioactive contamination to clean working zones. After use, the protective clothing is returned to the Hanford Laundry Facilities, operated by the Federal Support Services of ITT, for washing and decontamination. Prior to being returned to the user groups, the wearing apparel is monitored for alpha-beta-gamma contamination, and clothing is sorted for possible further contamination if necessary. Due to the large volume of protective clothing used by the various contractors throughout the Hanford complex, the monitoring process must be as rapid and efficient as possible.

This paper describes a monitoring system, which was developed by the Pacific Northwest Laboratory of the Battelle Memorial Institute to fulfill these requirements.

DISCUSSION

The developed protective clothing monitoring system incorporates a dual wire-belt conveyor to carry the miscellaneous types of protective apparel at a belt speed of 160 in./min between two large alpha-beta-gamma radiation detection probes. Eight, 5 in. diameter photomultiplier tubes are combined with eight, 6 x 7 in. light pipes and dual-purpose scintillators to provide 1, 2, or 4 fully independent monitoring channels approximately 24, 12, or 6 in. wide, respectively, and to accommodate the various widths of miscellaneous garments.

Incorporation of the solid state pulse amplifying and counting circuitry within the probe housings resulted in a low-noise and interference-free pulse counting system. Voltage outputs from the alpha and beta-gamma count rate circuits operate the high-level trip circuits that actuate the electromechanical rejection mechanism. Altering the high-level trip point is accomplished through adjustment of a potentiometer located on the front panel of the main instrument console.

Beta-gamma background, measured by a separate detector and count rate circuit, is used to automatically compensate the high trip levels. With the conveyor motor speed adjusted to drive the wire belt conveyors 160 in./min, a reliable high level contamination trip is obtained using a 25,000 dis/min (11 nCi) Ra DEF beta-gamma source or a 1000 dis/min (0.45 nCi) ^{239}Pu source distributed over a 100 cm^2 area.

The upper conveyor assembly can be raised to permit monitoring rubber overshoes. In routine operation, the belt speed is 160 in./min and the beta-gamma count rate circuit time constant is 3 sec. If increased sensitivity is desired, it can be achieved by reducing the belt speed and increasing the time constant of the count rate circuits.

An overall front view of the monitoring system is shown in Fig. 1. In monitoring miscellaneous protective clothing such as gloves, caps, etc., the upper conveyor belt is operated in contact with the lower belt. As the items of wearing apparel are loaded onto the conveyor, they are pressed between the upper and lower conveyor belts allowing the items being monitored to pass very close to the detectors to permit detection of alpha contamination. The stainless steel wire belts, manufactured commercially,* are 26.1 in. wide, constructed of 0.05 in. diameter wire, with a mesh size 0.5 x 2.0 in. This implies a wire belt open area of approximately 90%.

After the alpha-beta-gamma radiation is measured by the two scintillation probe assemblies, the item is conveyed to the rejection mechanism, shown in the rear view of the monitor, Fig. 2. The rejection mechanism consists of a linear actuator** with holding solenoid, coil tension spring, actuating arm, shaft, and wearing apparel slide. Contaminated items are deposited into a separate receptacle as a result of a trip signal causing release of the actuator-holding solenoid after ΔT-1.

DETECTION PROBE DESCRIPTION

Four 5 in. photomultiplier tubes (Amperex 54AVP) are used in each of the two large radiation detector

---

*Wire Belt Company of America, Winchester, Massachusetts

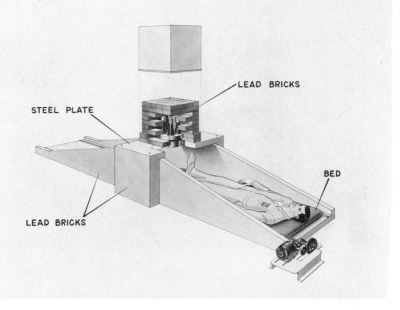

Fig. 1. Front View of System

Fig. 2.   Rear View of Monitor

assemblies and one 5 in. photomultiplier tube is used
in the background radiation detector. Separate Lucite
light pipes, 7 x 6 in. and 1 in. thick, were optically
coupled to each of the photomultiplier tubes. Beta-
gamma emitters are detected by a terphenyl-in-
polyvinyltoluene scintillator* 0.02 in. thick and cut
to cover the entire face of the light pipe. This thickness provides a nearly optimum ratio of beta-gamma
contamination response to gamma background. The
scintillator sheet is cemented to the face of the light
pipe using epoxy cement.** This provides excellent
optical coupling and a reliable bond.

The alpha detection scintillator*** is silver-
activated zinc sulphide in decal form, cut to size and
applied to the surface of the NE-102. The face of each
detector is covered with two layers of 0.2 mg/cm^2
aluminum "Dutch Leaf" and one layer of 0.9 mg/cm^2
double aluminized Mylar for light shielding.

Four of the described assemblies are placed side
by side in a light-tight metal container to form a composite radiation detector approximately 28 in. long
and 7 in. wide. The radiation accepting surface of
the detector is covered with copper household screen-
wire to protect the Mylar light shield. The two wearing apparel monitoring probes are mounted above and
below the conveyor belts, allowing simultaneous alpha-
beta-gamma monitoring of both sides of the item.

---

*Terphenyl-in-polyvinyltoluene (NE-102) manu-
factured by Nuclear Enterprises Ltd., Winnepeg,
Canada.

**NE-580 manufactured by Nuclear Enterprises
Ltd., Winnepeg, Canada.

***Har-D-Cal., Williston, South Carolina

A variable speed* motor maintains constant belt velocity and can be changed to other preset speeds as desired.

Details regarding circuitry utilized for pulse amplification, noise discrimination, count rate measurement, voltage trips, digital logic, and rejection mechanism operation may be obtained from the reference report.[1]

Time delays are necessary because the radiation detecting probes are located some distance from the discharge end of the monitoring system. The timing period of time delay ΔT-1 corresponds to the shortest possible time required for the item of wearing apparel to traverse from the probe to the discharge end of the conveyor belt. The period of time delay ΔT-2 corresponds to the longest possible time required for the contaminated item to pass from the belt to the "contaminated apparel" hamper. Time adjustments for time delays ΔT-1 and ΔT-2 were determined experimentally, based on the conveyor belt speed and the longest item to be monitored.

ITEM LOADING INDICATION & TOTALIZER

To properly set the minimum space between items, four sets of loading indicator lamps (photo-resistor controlled) are used. As an item is loaded on the conveyor belt, in any channel, it passes between the

---

*"Adjusto-Speed" induction motor, Model No. 5041 equipped with printed circuit board assembly No. 5-47 and speed reducer style A-2. Eaton Manufacturing Company, Kenosha, Wisconsin.

BLOCK DIAGRAM

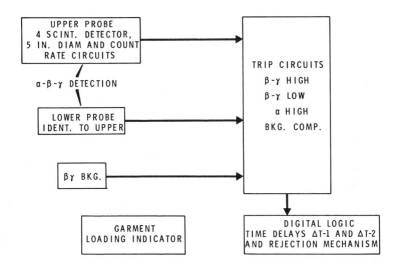

Fig. 3. System Block Diagram

photo-resistor and its light source. The resulting
voltage change activates the input transistor switch,
sets a binary, and energizes the red lamp. As the
item approaches the discharge end of the conveyor
belt, it passes between another photo-resistor and
light source assembly. This action also energizes
circuitry which de-energizes the red light, and a
green light is energized. The green light indicates
that another item may be placed on the conveyor.

An item totalizer, which operates from signals
received from the red loading lights, indicates the
number of items passed through the monitoring system.

For safety purposes, a stop switch (push-button)
is located to the left of the panel control switch (Fig. 1),
and an adjustable slip clutch is located in the drive
mechanism for the conveyor belt.

RESULTS AND CONCLUSIONS

Installed since February, 1967, the system has been well accepted by operating personnel with reliable operation achieved for approximately 7,000 hours of machine time, as of December, 1968.

Sensitivity to alpha and beta-gamma radiation exceeded requirements with 25,000 dis/min of Ra DEF and 1000 dis/min of 5 MeV alpha emitter (distributed over 100 cm^2 areas) reliably detected. Sensitivity can be readily increased, if desired, by decreasing belt speed and increasing the count rate circuits time constant.

Some minor mechanical difficulties were noted and corrected. Some rejected items were sufficiently stiff and failed to deflect downward, touched the bottom side of the slide, and move onto the lower conveyor assembly. This problem was solved by the addition of a soft-bristled wide brush between the item slide and the end of the conveyor belt.

To reduce frequency of light leaks, the original 1/2 in. square opening metal screening 80% open was replaced with 1/4 in. thus reducing the above alpha detection sensitivity from 1000 dis/min to 1500 dis/min. With this screening approximately one light leak per 1000 operating hours occurs. Wear necessitated replacement of the original wire belt at approximately 4000 hours of operating time.

Only two components have failed in the electronic circuitry.

Table I shows the number of various types of wearing apparel that can be monitored per 8 hr shift.

ACKNOWLEDGEMENTS

The authors thank C. A. Ratcliffe for his suggestions on portions of the electronic circuitry, and R. E. D. McGinnis for suggesting the addition of the soft-bristled brush and for maintenance records. The authors express appreciation to B. B. Evans of ITT Federal Support Services for his encouragement of the project.

Table I.

Wearing Apparel Items Typically Monitored Per 8 hr Shift

Item	Arbitrarily Selected Shift		
	1	2	3
Cloth and Rubber Gloves	1468	1220	2946
Cloth Caps	1274	none	none
Cloth Hoods	none	562	404
Canvas Shoe Covers	475	140	1015
Canvas Boots	1400	290	none
Towels	none	none	52
Rubber Overshoes	1371	1143	1048
Total Items Monitored Per Shift	5988	3355	5465

REFERENCE

1. E. M. Sheen and G. D. Crouch Automatic Conveyor Type Laundry Alpha-Beta-Gamma Contamination Monitor, BNWL-540, November 1967.

NEUTRON ACTIVATED CRITICALITY ALARM SYSTEM

L. V. Zuerner and R. C. Weddle
Battelle Memorial Institute
Pacific Northwest Laboratory
Richland, Washington

ABSTRACT

A system incorporating BF_3 type neutron detectors has been designed and is now in use at the Pacific Northwest Laboratory. The monitors have been calculated to alarm to a pulsed fast neutron dose of 50 micro-rem. They have been successfully tested under simulated criticality conditions with a 50 micro-second burst from a pulsed reactor. The alarm is activated when a preset count rate is exceeded. Alarm time is less than 0.5 seconds then the count rate---from neutrons exceeds the alarm set point counter rate---by 5 %. Basic sensitivity is determined by the type of BF_3 tube used ; the instrument can be adjusted to alarm in fast neutron fields of 20-1000 mrem/hr.

---

*Work performed under Contract Number AT(45-1)-1830 between the U. S. Atomic Energy Commission and Battelle Memorial Institute.

The signal from the polyethylene moderated BF_3 tube is amplified, shaped and fed to an integrating capacitor in a voltage sensitive trip circuit. This circuit activates the criticality alarms. Failure alarms are energized by circuits which detect high or low operating voltages and loss of signal from the neutron circuits. A self-auditing operating feature is designed around random input from background radiation.

The criticality alarms are operated in a "2 out of n system". Two monitors must be in the alarm condition before an evacuation alarm is sounded. This minimizes false alarms caused by local high level gamma fields, electromagnetic interference, and other spurious signals.

INTRODUCTION

In 1966, the Pacific Northwest Laboratory, Battelle Memorial Institute, undertook the development of a neutron sensitive criticality alarm. The original design, operation, and calibration of these units has been described by Friend.[1] This paper will discuss the original circuit designed by Friend, as well as subsequent modifications to improve reliability and usefulness of the monitoring unit. A 2 out of n matrix system, designed to eliminate false alarms from component failure and to provide maximum reliability, is also described.

OPERATION OF THE DETECTOR UNIT

A block diagram of the monitor is shown in Fig. 1. The output of the BF_3 tube averages about 575 counts per minute per mrem/hr for PuF_4 neutrons ($E_{AV} \sim$ 1 MeV). The operating point on the plateau varies from 1700 to 1800 volts, depending on the individual tubes.

The pulse amplifier is charge sensitive, with the input charge, per pulse, required to operate the one shot multivibrator and driver varying between 2 and 4×10^{-12} coulombs. The 2 to 4 microsecond, +15 volt pulses from the driver are converted in the log output circuit to a current to operate a 1 milliampere recorder. The recorder output versus mrem/hr is shown in Fig. 2. The driver output pulses are fed to the two trip circuits. These are voltage sensitive circuits using an integrating network for the pulses. When the count rate reaches a certain level, the voltage on the integrating capacitor is sufficient to fire a unijunction transistor which in turn fires an SCR which activates the alarm relays. The backup trip circuit is set to operate at a count rate slightly higher than the primary trip circuit.

These alarm circuits are count rate devices. When the count rate equals or exceeds the alarm set point, the alarm will be activated. The alarm time is less than 0.5 seconds when the count rate exceeds the alarm set point by 5 percent.

Tests, made during the modification program, showed that by varying components in the alarm circuit the trigger level could be varied from 12,000 to 560,000 counts per minute, corresponding to fast neutron dose rates from about 20 to 1,000 mrem/hr.

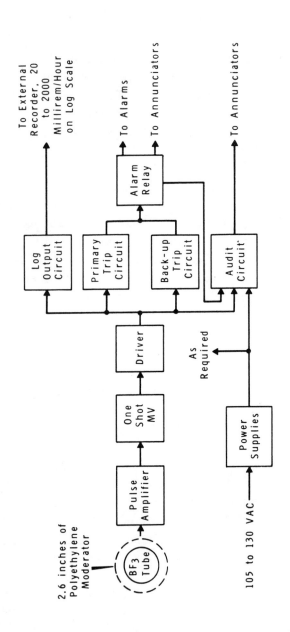

Fig. 1. Block Diagram of Neutron Sensitive Criticality Detector

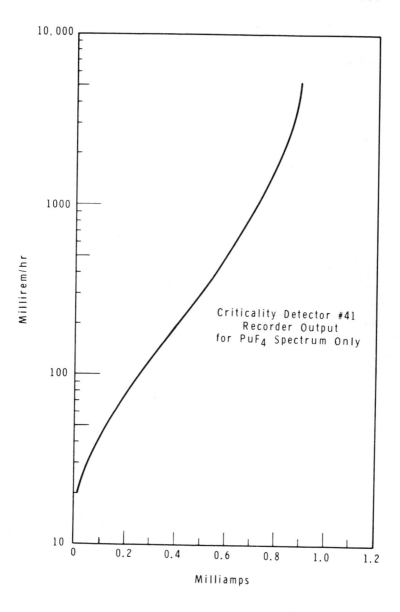

Fig. 2.   Typical Recorder Calibration

The audit circuit checks for and notifies if signals are not being received from the BF$_3$ tube or if the AC line voltage has dropped below the value for correct operation. If either of these two conditions exist, the yellow failure light will turn on and relay contacts will close providing an external signal, if one is desired. In the case of the former, the failure mode is activated if the count rate falls below the normal background from cosmic rays—about 3 counts per hour. This failure detection feature is highly desirable, and has proven successful in reducing manual checks.

The alarm relays are operated by signals from the trip circuit and audit circuit. When the counting rate from the BF$_3$ tube exceeds the trip level, relay K-1 closes. The blue criticality light on the front panel is energized and a set of contacts are closed to provide a signal for an external alarm. When the criticality light is on, the normal light goes off.

The audit circuit opens a relay in the alarm circuit in case of low AC line voltage or no signal from the BF$_3$ tube and related circuits. When this relay opens the normal light goes off, and the yellow failure light comes on.

MODIFICATIONS TO ORIGINAL DESIGN

Although the initial units were satisfactory for many situations, certain undesirable characteristics were noted: First, the instruments were susceptible to external electromagnetic interference, which could result in false alarms.[2] A second undesirable feature was the maximum high voltage, which was limited to about 1800 volts in the original units. This is not sufficient for many types of BF_3 tubes, thereby limiting the applications of the monitors. In addition, the high voltage could not be easily adjusted to the correct operating point on the BF_3 tube plateau, and the current through the regulator tube was approaching maximum rated values. Finally, the efficiency of the original BF_3 tube was too low. The small count rate per mrem/hr did not provide adequate signal to noise ratio at the trip circuit.

To provide for increased stability and wider operating range, the high voltage supply for the BF_3 tube was changed from the half wave rectifier type in the original design to a voltage doubler diode pump supply. The high voltage at the regulator tube may easily be varied from 1600 to 2400 volts by changing regulator tubes. Once the high voltage regulator has been selected, the high voltage at the BF_3 tube may be adjusted to a voltage which is between that of the regulator tube and 200 volts below that of the regulator tube.

In actual calibration, an external high voltage supply is connected to the neutron criticality detector. A scaler is connected to the collector of the univibrator transistor. In addition, a fast neutron source is used to provide the desired flux at the detector. The high voltage supply is then varied and a plot of high voltage versus count rate is determined. From this plateau the correct high voltage operating point is determined. The external supply is disconnected and the internal supply reconnected. The high voltage resistor is then adjusted to provide the correct operating voltage to the BF_3 tube. In this manner, the characteristics of the BF_3 tube are mated to the characteristics of the monitor.

In some instances a background and gamma plateau are also run, providing a more complete set of characteristics on the system is available. Thus BF_3 tubes with excessive background or gamma response will be found and can be replaced.

The small BF_3 tube used with the criticality detector gives an output of about 575 counts per minute per mrem per hour for fast neutrons from PuF_4. At dose rates of about 20 mrem/hr, the difference between the background and trip voltage in the trip circuit was 2 to 3 volts.

Outside electrical interference getting into the monitor and line transients would at times increase the voltage at the trip circuit to the alarm set point, causing a false alarm.

Experiments were made and a larger volume BF_3 tube was developed which would operate at the same voltage as the original tube. This new tube had an output of 2,000 c/m/mrem/hr. In a 15 mrem/hr field this count rate increased the voltage difference between the voltage needed for alarm trip and background voltage to about 10 volts. With the above arrangement, the alarm was set to trip in a fast neutron ($E_{AV} \sim 1$ MeV) field of 20 mrem/hr. The effect of outside electrical fields and line transients was reduced to almost zero.

TWO OUT OF N SYSTEM

To minimize false alarms from malfunction of a single detector unit, individual units were fitted into a two out of n system. In this system, several criticality detectors are used to monitor a given space.

In practice, the number of detectors is three or more, with placement such that if one fails, the space that it is monitoring is overlapped by one or more of the others.

In the case of the criticality alarm, two equal detection circuits are used in parallel. These circuits form a redundant comparator network using diode logic. When a single detector is tripped, its associated diodes place an opposing, but equal logical bias at the summing junctions. This elevates the junctions to near zero volts. When the second detector is tripped, the associated diodes place their logical bias at the summing junctions. The voltage at the summing junctions now increases in the positive direction and turns on gate diodes. The open gates allow current to flow to actuate relays. These relays close power circuits to operate the audible, howler type criticality alarm.

The annunciator panel lamps and photo relays, "the Sonalert" driver circuitry, and key lock switches, provide for alarm system check-out and removal of faulty detectors. If a detector fails either in the "critical" mode or the "failure" mode, the associated lamp on the annuciator panel will come on. Also, a neon lamp will energize a photo-resistor, which in turn, biases the gate diode to the on mode, activating a free running multivibrator which pulses the "Sonalert" on and off. This provides an audible warning signal at the control panel, alerting the building operator to initiate proper corrective action for the faulty detector. The key lock switches provide for the silencing of the "Sonalert" by opening the circuit between the photo relay and bridge biasing resistors in the "Sonalert" driver circuit. In addition, the key lock switches remove the faulty detectors from the circuit by breaking the inputs to the comparison diodes in the redundant detector comparator.

To further minimize any false trip, abnormal or criticality alarms due to AC line noise; filters were installed on each criticality detector. The AC line going into the detector is filtered and the output signals from the relays are filtered. All wiring is run in grounded metal conduits, with AC power obtained from isolation transformers connected to the emergency power system.

A six unit system has been in operation for several months, without a false trip, abnormal, or criticality alarm.

REFERENCES

1. P. C. Friend, "Technical Manual for Neutron Sensitive Criticality Detector," BNWL-MA-29, May 16, 1966.

2. D. H. Denham, and J. M. Selby, "Criticality Detection System—A Plea for Current Criteria," BNWL-SA-2101 (1968); Presented before Health Physics Society Midyear Symposium, Los Angeles, 1969.

ACKNOWLEDGMENTS

The authors are indebted to P. C. Friend for his helpful comments, and to E. M. Sheen for redesign of the high voltage circuit.

DIFFERENT TYPE GLOVE BOX
FILTERS USED FOR ^{238}PU WORK

K. C. Angel and H. F. Anderson
Monsanto Research Corporation
Mound Laboratory*
Miamisburg, Ohio

ABSTRACT

A review is presented of the different type glove
box filters that have been used with ^{238}Pu glove boxes
at Mound Laboratory. Three filter installation types
are discussed: (1) box-type filter mounted inside a
small housing on top of glove box, (2) cylindrical
filter with 2-in. diameter threaded pipe at one end
designed to screw into exhaust line inside glove box,
and (3) dual cylindrical filters with molded rubber
gaskets designed to fit inside a stainless steel or
fiberglass tube below glove box. Each type filter installation is discussed in terms of filtration adequacy,
sources of leaks, ease of filter change, contamination

*Mound Laboratory is operated by Monsanto
Research Corporation for the U. S. Atomic Energy
Commission under Contract No. AT-33-1-GEN-53.

control problems during filter change, and interference with glove box operations.

INTRODUCTION

Mound Laboratory, Miamisburg, Ohio, has 10
yr experience in large-scale glove-box operations with
plutonium-238. The relatively high specific activity
of this isotope, approximately 17 Ci/g, plus its high
radiotoxicity have presented serious problems in
containment and in the prevention of internal exposures to personnel.

One feature of glove-box design which could be
improved from the applied health physics point of
view is the primary glove-box filter installation at the
exhaust of the glove box.

One of the primary filter installations used on
plutonium-238 glove boxes at Mound Laboratory was
a box-type housing mounted external to each glove
box. An improved design for general use is a cylindrical filter housing mounted inside the glove box.
Recently, a completely new design was placed into use
when the Plutonium Processing Facility at Mound
Laboratory went into operation in March, 1968. In
this latest design two cylindrical filters in series
are mounted in a stainless steel or fiber-glass exhaust tube below the rear of the glove box.

In the following sections the different type primary filter installations used at Mound Laboratory are
described and the major lessons learned by experience
with each type are reviewed.

DESCRIPTION OF PRIMARY, GLOVE-BOX FILTER INSTALLATIONS

A. Box-Type Housing, External to Glove Box

An example of the box-type filter installation external to the glove box is shown in Fig. 1. Four separate filter housings and part of a fifth may be seen in this photograph. Each filter housing serves the exhaust from one glove box. The housing has an upper and a lower chamber, each containing one absolute filter. This provides two filters in series at the exhaust of the glove box. There are four doors on each housing, two in the front and two in the rear. The rear doors are not visible in the photograph.

The absolute filter used in this installation is 8 in. x 8 in. x 5-7/8 in. with a plywood frame, gasketed on both faces. The filter is rated for 50 ft^3/min. The gaskets are compressed by sliding the filter into the chamber. The doors on the housing are gasketed and held shut by latches.

B. Cylindrical Housing, Mounted Inside Glove Box

An example of the cylindrical filter housing which is mounted inside the glove box is shown in Fig. 2. The filter housing is a steel cannister 6 5/8 in. in diameter and 9 1/2 in. long. The outlet end of the housing screws directly into the exhaust line of the glove box. The inlet end is fitted with a small holder containing a roughing filter.

Inside the cannister are two absolute filters in series. Each internal filter is 6 1/2 in. in diameter

Fig. 1. Box-Type Housing, External to Glove Box.

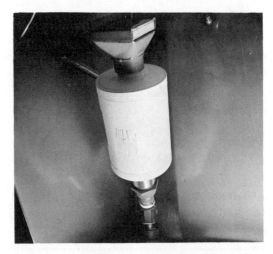

Fig. 2. Cylindrical Housing, Mounted Inside Glove Box.

and 2 13/16 in. long.   The complete assembly is rated
for 25 ft^3/min.

There is no gasket associated with this type installation.   The seal is dependent upon screwing the
threaded pipe tightly into the exhaust line.   Teflon
tape is normally applied to the threads; however,
care must be taken to minimize the amount of tape
which will be in contact with high level contamination
because (a, n) reactions with the fluorine in the Teflon
can result in a significant neutron yield.

C.   New, Primary Filter Installation

The newest type of primary glove box filter installation at Mound Laboratory is shown in Figs. 3
and 4.   Figure 3 is a photograph of the rear of a glove
box in the new Plutonium Processing Facility, showing the access to the filter tube.   Figure 4 shows a
closeup of the top of the filter tube with the upper
filter in position.   The screen, coarse fiber material,
and cap visible in this picture are placed over the
top of the filter tube to serve as a roughing filter and
to protect the face of the upper filter.   Figure 5 shows
one of the two absolute filters which are positioned in
the tube, one behind the other.   The filter itself is
5 1/4 in. in diameter and 6 in. long.   The gasket
around the top of the filter is a dual lip made of
molded silicone rubber which is compressed when the
filter is inserted into the exhaust tube.   Three wooden
spacers are attached near the bottom of the filter to
keep the filter positioned straight inside the tube.   The
filter is rated at 30 ft^3/min.

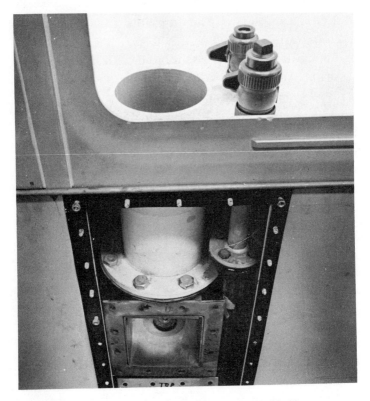

Fig. 3. New, Primary Filter Installation.
Rear of Glove Box Showing Access to Filter
Tube.

Fig. 4. New, Primary Filter Installation. Top
of Filter Tube With Upper Filter in Position.

Fig. 5. One of the Two Absolute Filters Which
are Positioned in the Tube.

EXPERIENCE WITH DIFFERENT
TYPE GLOVE BOX FILTER INSTALLATIONS

A. Box-Type Housing, External to Glove Box

1. Filtration. This type filter installation has proven very satisfactory in filtering the air exhausted from a glove box. Although such installations have never been tested "in place" by the D. O. P. technique, air samples collected downstream from the filters have confirmed that the installation does clean the airstream with a high degree of efficiency.

2. Source of Leaks. If the exhaust air from the glove box contains acid fumes, corrosion of the metal filter housing may occur over a period of time. In some installations there may be enough damage sustained from acid corrosion that leaks develop around the gasket seal.

3. Ease of Filter Change. As can be observed from Fig. 1, access to these filter installations above the glove boxes for a filter change is awkward. Personnel changing filters must work on ladders and reach out to the filter housing. In some laboratories, the space above the glove boxes is even more cluttered with pipe lines and equipment than is evident in Fig. 1. Some difficulty has been experienced in pushing the old filter out during the change procedure, but this is not normally the case.

4. Contamination Control During Filter Change. Large plastic bags are sealed over the front and rear doors to permit bagging both the old filter out and the

new filter in. Because of the awkward working conditions and the high level of contamination normally present on the lower filter, changing these filters has been a major contamination control problem. The fact that the filter housings are located out in the open laboratory could give some difficulties, since a large area is subject to contamination if anything goes wrong. A number of releases of activity have occurred at Mound Laboratory during filter changes with this type installation on plutonium-238 box lines.

This experience with contamination control during filter changes was the main reason that Mound Laboratory largely abandoned this type filter installation on plutonium-238 glove boxes.

5. Interference With Glove-Box Operations. The external type filter housing takes up no valuable space inside the glove box. It must be recognized, however, that normal operations in a glove-box line equipped with this type installation are interrupted if any contamination results from the filter change. As indicated above, releases of activity may occur during filter changes with this type installation and this, of course, results in some shutdown of glove-box operations during decontamination.

6. Other Considerations. The processing of large quantities of plutonium-238 dioxide does pose an external radiation control problem. The (a, n) yield of plutonium-238 dioxide is on the order of 4×10^4 n/sec/g. Some glove-box lines have to be shielded with several inches of a neutron shield material. The accumulation of plutonium dioxide in the unshielded external filter housings along a box line raises the dose rate in the general work area.

B.   Cylindrical Housing, Mounted Inside Glove Box

1.   Filtration.   This type filter installation is very
satisfactory, provided it is properly installed and does
not leak.   As in the case of the box-type housing, these
cylindrical housings have never been tested "in place"
using the D. O. P. technique, but air samples downstream from these installations have demonstrated
that they can clean the air with great efficiency.   Some
trouble has been experienced with leaks in this type
installation.

2.   Source of Leaks.   With very few exceptions, the
leak occurs at the pipe threads where the cannister
screws into the glove-box exhaust line.   A leak at this
point permits some of the glove-box atmosphere to be
drawn directly into the exhaust line, bypassing the
filters.   The problem here is that the threaded fittings
of the exhaust line become corroded and when the
primary filters are changed, the new cannister cannot
be, or simply is not, screwed tightly enough into
position to achieve a good seal.   To correct this
problem,   the metal fitting of the exhaust line has been
replaced in some cases with a PVC threaded fitting.

Another source of leaks may be the corrosion of
the cannister itself.   A hole in the cannister after the
last filter will also allow air to bypass the filters.   In
a few installations where this has been a problem, the
regular cannister has been replaced with one having a
stainless steel or PVC casing.   This has prevented
corrosion holes in the cannister itself, but makes recovery of plutonium dioxide from these cannisters
more difficult.

3. ·Ease of Filter Change. As indicated, corrosion
in exhaust line threads may make the physical act of
changing the filter difficult, and this increases the
chance that the cannister will not be screwed in
properly.

It should be mentioned that since this type installation takes up space inside the box, it is normally
located in the rear or off to one side so as to interfere as little as possible with the normal use of the
box. Because of their location inside the box, these
cannisters cannot be changed by using the main gloves
and the box must be equipped with a set of upper glove
ports in the viewing window.

4. Contamination Control During Filter Change. All
filter changes take place inside a glove box.

5. Interference with Glove Box Operations. This type
filter installation uses up about one cubic foot of
valuable glove box volume. The change of filters in
one glove box does not interfere with other glove-box
operations in the laboratory.

6. Other Considerations. A buildup of material on
the filters does increase the dose rate at the operators position; however, as these boxes are shielded,
this problem is not too serious.

C. New Cylindrical Filters, Installed In Exhaust Tube

1. Filtration. Before the new Plutonium Processing
Facility became operational in March 1968, approximately 30% of the glove-box exhaust filter installations

were selected at random and D. O. P. tested. With
two exceptions, no penetration could be detected.
This, of course, was with two absolute filters in
series, each of which is rated at least 99.95% efficient.
The lower limit of detectability in this D. O. P. test
was 0.001% penetration. The two exceptions which
showed some leakage were found to have defects in
the exhaust tubes. These exhaust tubes were out-of-
round so that the gaskets on the filters did not seat
properly.

Since the facility became operational, air samples
collected from the glove box exhaust system has confirmed that this type installation does provide excellent filtration.

2. Source of Leaks. There have been some glove
boxes where high level contamination has been noted
in the chamber below the exhaust tube. In these cases,
these glove boxes had been used for processes where
a large quantity of acide fumes were generated. The
Plutonium Processing Facility does have a separate
acid fume exhaust system which is designed to collect
the fumes locally inside the box and prevent them from
being carried into the main glove box exhaust. In
these isolated cases, apparently large volumes of
acid fumes were carried to the glove-box filter installation. It is not known how the material penetrated
the filters; however, one possibility is that the fumes
condensed above the filter gasket, and liquid migrated
past the seals to the lower chamber.

3. Ease of Filter Change. The changing of filters in
this type installation is much easier than in either of
the other two cases described earlier. Refer to Fig. 3.
As may be seen in this photograph, there is a cubical

chamber directly below the exhaust tube. This chamber
is opened by removing a bolted, gasketed plate. Once
opened, a new filter is placed in and pushed upward
into the exhaust tube. Normally, the filter is forced
upward into the tube with the aid of a small jack. As
the new filter is forced upward in the tube, the top
filter is forced out into the box and the middle filter
moves up to the top position.

4. Contamination Control During Filter Change. The
design of this installation lends itself to good contamination control. The filter change is made in a controlled
access service corridor behind the glove-box line.
The removal of the cover plate over the chamber below the exhaust tube is carefully monitored. If any
contamination is detected, a small plastic enclosure
with glove ports is used to complete the job. To date,
over 90% of these filter changes have been made without any significant contamination being detected.

5. Interference With Glove-Box Operations. The
protrusion of the exhaust tube into the glove box and
the need for room to push the dirty filter out during a
change uses up perhaps one-half of a cubic foot of
glove-box volume.
      The change of filters on one glove box does not
interfere with other glove-box operations in the laboratory.

6. Other Considerations. As in the case of the internal cannister type glove-box filter, accumulation of
material on the filter does increase the dose rate at
the operators position, but the filters are inside the
shielded box so that the problem is less serious than
in the case of the external filters.

THE USE OF A DIFFERENTIAL VOLTAGE COMPARATOR AS AN ALARM LEVEL CONTROL*

John Chester
Health Physics Division
Brookhaven National Laboratory
Upton, New York

ABSTRACT

The purpose of most fixed radiation monitoring instruments is to provide an alarm when the radiation levels being monitored exceed a predetermined value. It is therefore necessary to provide reliable alarm circuitry if the primary instrument is to effective.

Although meter relays are the most common alarm level control now in use they have several limitations. Contact type relays prevent the indicating meter from exceeding the alarm set point and they are limited to only one up-scale point and one down-scale point. The

*Research carried out at Brookhaven National Laboratory under contract with U. S. Atomic Energy Commission.

non-contact type meter relay is expensive and in both types the alarm level can only be set at the indicating instrument.

The disadvantages of the meter relay can be overcome by using a differential comparator in the form of an integrated circuit which is small, reliable and inexpensive. It can be connected to operate with either a positive or negative signal.

The alarm level control can be set accurately and is stable over long periods of time. The control can be located on the primary indicating instrument or at any remote location. The circuit requires so little power from the indicating instrument that several alarm points can be used. The circuit is wired in a fail-safe manner in that any failure of the indicating instrument, the differential comparator, or any other component will result in an instrument failure alarm.

Alarm circuits using Fairchild μA 710 differential comparators have been in use at Brookhaven's High Flux Beam Reactor air monitoring system for over a year without failure.

INTRODUCTION

The primary purpose of a radiation monitoring system is to provide an alarm when the radiation level exceeds a pre-set value. A common type of alarm level control is a meter relay equipped with one or two movable contacts which are closed by the pointer when the alarm level is reached. One disadvantage of a meter relay is that the number of alarm points is limited to two. Also, the pointer cannot indicate a level beyond the alarm set point. Several modifications on the basic meter relay have been developed, including those which use an oscillator or beam of light to improve the characteristics of this type of alarm level control. All meter relays have the inherent limitations of a mechanical device. They are subject to wear and friction which causes sticking and prevents them from being fail-safe.

Reliability is a primary criterion in designing any electronic circuit, especially an alarm circuit. The limit of the circuit's reliability is determined by the components which make up the circuit. In the past few years, integrated circuits have been developed which have very stable DC characteristics over a wide temperature range and response times which are superior to comparable discreet circuit components.

THE BASIC CIRCUIT

An alarm circuit has been designed using the Fairchild μA 710 high speed differential comparator. A differential comparator is a high-gain differential input, single-ended output amplifier. The function of the circuit is to compare a signal voltage on one input with a reference voltage on the other and produce a change in the output level when one of the inputs becomes more positive than the other. One input functions as an inverting input and the other as a non-inverting input. When the non-inverting or positive input is positive with respect to the other input, the output is positive. When the negative or inverting input is positive with respect to the other input, the output is zero.

The μA 710 will respond to a voltage difference at the input as low as .6 mV and is designed to operate with input signals as high as $\pm$ 5 volts. The absolute maximum rating is $\pm$ 7 volts and should not be exceeded. Although the input voltage range of the circuit is $\pm$ 5 volts, the maximum voltage difference between the inputs is 5 volts. Therefore, if one of the inputs is at + 5 volts, the other can only be driven to as low as ground without exceeding the differential input voltage limit.

In this circuit the differential comparator is
being used as a low hysteresis variable threshold
Schmitt trigger. A comparator is similar to a dif-
ferential-input operational amplifier except that it is
expected to recover quickly from saturation.

When the circuit is used as an alarm level control,
one input is connected to the output of the measuring
instrument and the other is connected to an adjustable
reference voltage. The potentiometer which sets the
reference voltage is the alarm level control. If the
maximum voltage swing of the potentiometer is equal
to the maximum signal voltage available at full scale
deflection, the alarm potentiometer is calibrated in
percent of full scale. The signal voltage may be
connected to either the inverting (-) or non-inverting
(+) inputs which makes the circuit applicable for use
with instruments that supply either a positive or
negative output signal. The most convenient output
signal to use is normally the 1 mA recorder output.
If a recorder is used, the circuit may be connected
directly across it or, if no recorder is used, a dummy
load must be inserted.

Figure 1 shows the connection to the comparator
circuit for a positive input. The input is connected to
the inverting or negative input and the alarm level
control which supplies the reference voltage is connected to the non-inverting or positive input. When
the input signal is less positive than the reference
voltage, that is, below the alarm set point, the output of the comparator will be positive. In order to
insure the stability of the circuit at the alarm point,
some hysteresis has been designed into it. Resistor
R_2 provides a small positive voltage at the reference
input while the output is positive. This voltage is

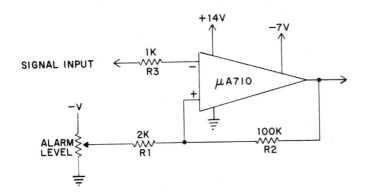

Figure 1.
Connection to the comparator circuit for a positive
input.

determined by the ratio of R_1 to R_2. In this circuit,
when the signal input is below the reference, the output is positive, adding a small voltage to the reference
input. When the signal level rises above the alarm set
point, the circuit trips and the output goes to zero.
The voltage which was added to the reference is removed, increasing the difference in voltage between
the two inputs. This decrease in the reference voltage
has the same effect on the comparator as an increase
in the alarm signal and produces sufficient hysteresis
to insure that the circuit will not oscillate around the
alarm point.

FAIL SAFE OPERATION

Figure 2 shows the circuit complete for "fail-
safe" operation. With the input signal less positive
than the reference, the output of the comparator is
positive causing the NPN transistor, T_1, to conduct
holding relay R_1 closed. When the signal goes more

positive than the reference or above the alarm set point, the output of the comparator will drop to zero. As the base of T_1 goes negative it turns off, allowing the relay to open. The alarm circuit is wired to the normally closed contacts of the relay; therefore, when the relay opens, the alarm contacts are closed. The failure of any component, or of the power supply, will produce an alarm which makes the circuit "fail-safe." Except for the action of the relay, there are no moving parts such as there are in a meter relay circuit. The relay is wired in series with one set of its own normally open contacts so that it will not reset itself when the radiation level drops below the alarm set point. A normally open push switch is wired across the relay contacts to facilitate resetting the alarm circuit. This latch-out feature can be omitted by wiring the relay directly to the collector of the transistor; the alarm signal will then stop when the condition causing the alarm no longer exists.

In the circuits described here, a switch is provided at the input to test the operation of the circuit by applying a voltage to the input equal to the full scale signal voltage. The function of the alarm set control is tested by making the measuring instrument read a value equal to the alarm set point value.

The input of the circuit is protected against over-voltage by resister R_3 and Zener diode D_1. The value of the Zener diode is selected to be slightly higher than the maximum signal voltage (5.6 volts). If a high potential is accidentally placed across the input, the diode will start to conduct. Resister R_3 acts as a current limiter and will burn out if the overvoltage is high enough, protecting the comparator. Resister R_4 maintains a load on the input when the circuit has been

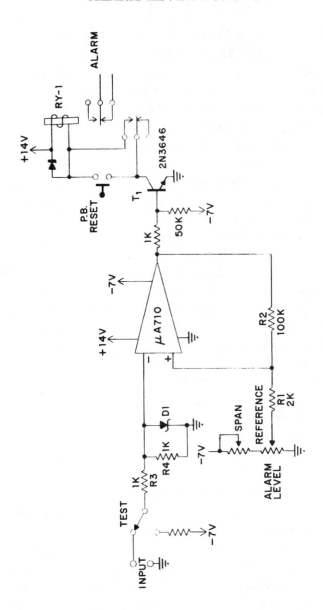

Figure 2.

The complete circuit equiped for fail-safe operation

disconnected from the measuring instrument.

MULTIPLE COMPARATORS

A second comparator circuit may be added (Fig. 3) to act as a low level or instrument failure alarm. The input is connected to the non-inverting or positive input and the reference is connected to a positive supply. Normal background radiation will usually cause the measuring instrument to indicate some level above zero. In cases where this is not true, it is sometimes advisable to install a small source on the detector. The effect of the source will test the function of the detector and measuring instrument. The reference is set just below this level so that during the normal operating condition the signal is more positive than the reference causing the output to be positive, holding the relay closed. If the instrument fails and the signal drops to zero, below the reference voltage, the output of the comparator will go to zero which allows the relay to open causing the alarm. The operation of this circuit is also fail-safe and is tested by shorting the input to ground which drops the input below the alarm set point.

Any number of circuits can be added with their own set points so that intermediate conditions can be monitored.

AVAILABILITY AND RELIABILITY

The μA 710 comparator is an integrated circuit which makes use of low wiring capacitance and excellent component matching available with monolithic construction. The circuit which is etched on a 35-mil-square piece of silicon, contains 10 transistors plus the associated diodes and resistors. It is fabricated in either a flat pack design or a TO-99 can with 8 leads and is readily available at about $5.00, which makes the price of the entire alarm circuit less than a meter relay. A μA 711, which is a dual comparator, that is

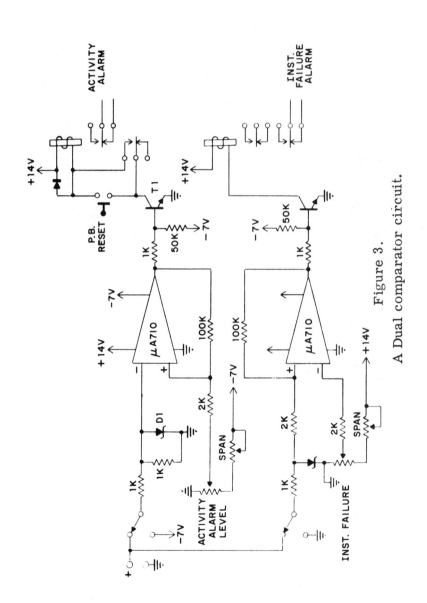

Figure 3.

A Dual comparator circuit.

two comparator circuits in one can, is now available
which would make the inclusion of the low level alarm
even more compact. The relay used a low current
reed relay with DPDT contacts rated at 10 volts/amp
at 1 amp. They are epoxy-resin encapsulated with
pin-type terminals for mounting on printed circuit
boards. The relays may also be replaced by silicon-
controlled rectifiers if the "fail-safe" or latch-out
feature is not desired. Electrically, the circuit requires a +14 and -7 volt power supply at approximately
5 mA each. All of the components have been mounted
on a printed circuit board designed as a plug-in unit.
The circuits described here have been in service at
the Brookhaven National Laboratory's High Flux Beam
Reactor for over a year without a component failure.

Session VI
AIR MONITORING

Chairmen

SEYMOUR BLOCK
Lawrence Radiation Laboratory
Livermore, Calif.

PAUL O. MATHEWS
Sandia Corporation
Livermore, Calif.

AIR SAMPLING IN OPERATIONAL MONITORING*

Harry F. Schulte
Health Division
University of California
Los Alamos Scientific Laboratory
Los Alamos, New Mexico

The subject of air monitoring still needs treatment in a monograph. In a paper like this, which is intended to initiate a brief symposium on the subject, one can only touch lightly on a wide variety of aspects. In doing so, one raises many more questions than he answers, but hopes to stimulate interest in these questions. Air monitoring as a technique seems deceptively simple. But this illusion is dispelled as soon as one asks the first obvious question—"When is air sampling required?" Routine air monitoring is really required only in installations handling large quantities of radioactive material. This means

*Work performed under the auspices of the U. S. Atomic Energy Commission.

something above the usual laboratory scale and may include the following:

1. Tritium in large scale production processes and heavy water reactors and iodine isotopes at levels of a few hundred millicuries.

2. Reactor fuel fabrication and reprocessing and machining of natural and enriched uranium

3. Processing of plutonium and other trans-uranium elements; and

4. Uranium mining, milling, and refining.

These suggestions are quoted from ICRP Publication #12 on General Principles of Monitoring.[1] Much air sampling is done on other processes where good house-keeping will prevent air contamination and surface monitoring and other procedures will indicate its absence.

The next question one should ask before designing an air monitoring program is "What are the purposes of sampling?" One can cite many reasons in specific cases, but there are two predominant ones:

1. For the evaluation of the effectiveness of en-vironmental control measures and equipment

2. For estimating the probable upper limit of inhalation of radioactive materials by workers.

The first reason is most important and air sampling data is a very valuable tool in defining the steps needed to reduce air contamination. Thus air sampling is essentially a tool for operational monitoring in areas where significant quantities of radioactive material can become airborne.

Specific types of air sampling equipment have been described in numerous books and papers. It is impossible to say that a particular instrument is best for all possible situations. Much sampling equipment is fabricated by the user to meet his specific needs. Good commercial equipment is available, but even this may require some modification to meet particular requirements. Practically all sampling for particulate matter is done by filtration. In most cases the filter must be removed and counted after completion of the sampling period but there are many ingenious devices available for counting the activity while it is being deposited. This is relatively simple for beta and gamma emitters, but more difficult for alpha emitters. The presence of naturally occurring radioactivity is always a problem and is handled by delayed counting, alpha-beta coincidence counting, alpha-to-beta ratio measurement, energy discrimination, or other methods.

The type of filter paper to be used also cannot be simply specified because requirements vary with circumstances. The paper must have a fairly high collection efficiency for the material to be collected yet it must offer low enough air resistance so the required flow rate can be maintained without serious change during the sampling period. For alpha emitters, collection should occur near the surface to minimize losses during counting. There are many papers available that meet these requirements[2] and the choice is not a critical one. Instead of individual sampling units each operating on its own pump, many facilities now use central sampling systems where numerous filter heads operate from a large central pump.

Measuring gaseous contamination of the atmosphere is more varied since there is no common collection technique similar to filtration. Iodine is
sampled by drawing air through a cartridge of activated
charcoal and then gamma-counting the cartridge.[3]
The cartridge should be preceded by a filter that is
also counted because some of the iodine occurs in
particulate form. We have not found carbon or silver
impregnated filter paper satisfactory collection media
for iodine.[3, 4]

The noble gases, argon, krypton, or xenon, are
usually determined by a beta or gamma detector
directly in the atmosphere to be monitored. This may
be a Geiger counter or an ion chamber through which
the air is drawn as in the Kanne chamber.[5] These
gases can also be collected on charcoal but the charcoal must be kept very cold.

Tritium and tritium oxide vapor are commonly
measured by passing the air through an ion chamber.
Particulate matter is usually removed first by a filter
and existing ions are removed by an electrical ion
trap. The instrument is more sensitive and less
accurate if existing ions are not removed, but this
cannot be used if sources of ionization other than
tritium are present.

Radon is determined by collection in chambers
lined with scintillating material that is subsequently
counted on a photomultiplier. A flow-through ionization chamber may also be used as well as collection
on refrigerated activated charcoal. In uranium mining
the determination of randon is of much less interest
than measurement of the concentration of radon decay
products. These are short lived particulates and are
collected on filter papers. Because of their short

half-lives, the length of sampling and the time between sampling and counting must be measured and regulated. The concentration is then determined from calibration curves and expressed in terms of "working level" without distinguishing between the various nuclides present. [6]

These few paragraphs are all that can be said here on the subject of specific equipment for air sampling. These subjects are very well covered in the literature. When one looks at the problem of evaluation and control of airborne hazards, the real practical difficulties lie not in the equipment or its operation, but in its applications and interpretation of results. The rest of this paper will deal with these problems as related to operational monitoring.

It is useful to distinguish three types of situations under which sampling may be used:

1. Repetitive type operations. These are industrial type operations where the worker repeatedly carries out a fixed series of operations throughout the working day and on successive days. Accurate measurements of air concentrations during each operation can lead to a close estimate of the worker's daily weighted average exposure and can identify the phases of the operations contributing to air contamination.

2. Irregular activities of a non repetitive nature. These are characteristic of scientists in laboratories. Here no series of operations is representative of any average conditions. If sampling is required, it must be continuous.

3. Constant work with enclosed processes. Plutonium or fuel element processing is an example. Routine air concentrations are usually very low except when there is a failure of control equipment such as glove box leaks, spills, of similar incidents. Sampling usually is continuous and a sampler which gives an alarm at high air concentration levels is necessary.

Proper location of samplers is extremely important. For repetitive operations, the sampler may be hand-held by the health physicist in the worker's breathing zone or the worker may wear a personal air sampler. For irregular activity, fixed or installed samplers must be carefully placed in various working locations, especially where air contamination seems most likely and where people work most frequently. This can be done best by close observation and experiment. Personal air samplers should be used in a study to determine how accurately the fixed samplers assess the actual exposure of the workers. For enclosed processes, a careful study must be made of airflow patterns in the workroom. The alarm sampler should be placed where it responds most quickly to releases of radioactive material anywhere in the room. The point where the general room air leaves the room will often be the best location. Fixed samplers in the workroom should be located at positions of likely contaminant release. The easy solution to questions of sampler location is to increase the number of samplers in use, but this easy solution will create more difficulties in interpretation because interpretation depends on a knowledge of how much time the worker spends in each location. Although the work may not be repetitive it should be possible to make rough estimates of exposure duration based on observations and time studies.

That the size of particulates in air is important in evaluating air monitoring data is well recognized. The report of the Task Group on Lung Dynamics offers a means of relating particle size to deposition in various parts of the respiratory tract.[7] Following our usual propensity for mathematical manipulations, a number of health physicists have developed methods of calculating tissue doses from size and concentration parameters.[8, 9, 10] This is interesting and useful but there is still a great need for actual size measurements under various working conditions.[11, 12, 13, 14, 15] As so frequently happens following similar calculations, the large degree of uncertainty involved in the measurements and assumptions may be forgotten and a higher degree of exactness is attached to the final result than it deserves. Nevertheless, the calculations have been valuable as estimates of the degree of hazard associated with airborne contamination.

Another approach to the particle size problem is the use of two-stage air samplers in an attempt to distinguish between material which will deposit in the upper and lower respiratory tract. This approach is now being used on nonradioactive materials such as coal and silica dust and deserves more consideration by health physicists.[16, 17, 18] The ICRP model requires a determination of activity median aerodynamic diameter which is not given by two-stage samplers. However, pulmonary deposition as predicted by the ICRP model is approximately related by a simple

factor to the quantity collected on the second stage of
a sampler whose calibration follows the standard AEC
curve.[19, 20] Graded filter papers also offer a means
of directly relating physical measurements to respiratory deposition.[10, 21] This method may become more
useful than any present technique.

Regardless of the method used, there is an urgent
need for more measurements of the particle sizes of
materials for which we are doing air sampling. The
various cascade impactors and the cascade centripeter
are adequate instruments for measurements of the
full particle size spectrum. Nuclear track measurements are a useful adjunct for alpha emitters and are
important in operational monitoring, not only in dose
estimation.

In addition to size distribution, it is desirable to
measure the activity distribution of airborne material.
Radioautography is the principal tool for this purpose.
In many cases it has been found that a high percentage
of the activity on a sample that has been run all day
is concentrated in a single particle.[14, 22] This
obviously affects the way in which the result is interpreted with respect to dose. A single hot particle
may be large and hence not reach the lower respiratory
tract. Even if the particle is small, the health problem which it presents is not one of general irradiation of a large organ such as the lung, which is the
basis for our present concentration standards. Recently a model has been proposed for calculating the
probability of tumor formation from individual
particles.[23] If most of the activity collected by a
sampler in the course of a day is on a single particle,
the probability of inhaling such a particle should be
assessed. Most important of all, the frequent

appearance of high activity particles tells us something about the airborne contamination that may identify the source of this contamination and lead to its
elimination. Although only a small fraction of air
samples can be studied for activity distribution, such
analyses should be done frequently for the information
they yield.

The factor of solubility is also important and is
difficult to assess. The present MPC for soluble
plutonium is one-twentieth of that for insoluble
plutonium, which is a very significant difference. The
proposed ICRP lung model would indicate even greater
differences between extreme transportability classes.
Differences are not only simple solubility factors but
must include chemical differences due to formation of
colloids, complexes, and chelates. Practically
nothing is being done to study these differences among
airborne materials in the working environment.
Caldwell[24] has suggested a field test which does distinguish certain solubility differences and we need
many data of this sort. We may not always be able to
assume the maximum or minimum degree of solubility,
whichever is the most serious risk. This is especially
true where differences are great or where one risk
may be of irradiation of bone and the other that of a
lung tumor. Solubility differences also help to define
sources of contamination.

The way in which airborne concentration varies
with time is a useful piece of information. This may
require no additional measurements, but only a study
of past results. A control chart type of plot or simple
examination of past records may reveal trends indicating conditions that need correction. If concentrations
at a given location are plotted against the frequency

A Few Typical Particle Size Measurements

	Mass Median Diameter	Standard Deviation	Source
Plutonium			
Fluorination	.45 m	1.5	Los Alamos
Fluoride reduction	.30	1.5	Los Alamos
Lathe operation	.30	1.4	Los Alamos
Burning metal	.10	1.5	Los Alamos
Pressurized suit area	5 – 10		Harwell
Glove box area	5 – 10	2	Harwell
Powder metallurgy	1.0		Harwell
PuO$_2$ hood	1.5	2.5	Hanford
PuC processing	0.6	1.5	Hanford
Uranium			
Foundry	1.9 – 3.2 m	2.1 – 3.7	Los Alamos
Machining	0.4 – 3.9	2 – 5	Los Alamos
Extruding	3	2.5	Los Alamos
Screening oxide	1.0 – 3.5	2 – 3	Los Alamos

of a given concentration level, a logarithmic probability distribution is usually found. The geometric standard deviation of these results, easily obtained from a plot of the data on logarithmic probability paper, is a measure of the variability of concentration. By comparing such results from one location with those of another, additional information is gained which is useful in locating the source of airborne contamination. A high variability also means that only occasional sampling at that location may give a very incorrect estimate of the average concentration.[22, 25]

In addition to variation of concentration with time, one should consider variation in space. It is obvious that the air contaminant is not evenly distributed throughout a workroom. However, it may not be appreciated that concentrations may vary greatly over a distance of only a few inches. We have run identical samples 12 inches apart for long periods and obtained results differing by a factor of 3. Other investigators have equipped workers with two personal air samplers on opposite sides of the front of their bodies and obtained widely varying results.[26, 27] The significance of this is that under the best of circumstances an air sample at a given location is only a rough estimate of the average concentration in a volume a few feet in diameter around the sampler. Even a personal sampler does not give an exact measure of the concentration breathed. The distribution of air concentration throughout the space in a workroom is important in locating sources of contamination. For this purpose air monitor results must be correlated with studies of airflow in the room, and individual samples in a workroom should be studied for correlation with each other.

Most of my preceding remarks may seem to cast doubt on the accuracy of conventional air sampling as a measure of individual exposure, and this is a fair conclusion. [20] Air samples taken in fixed locations can give only a very rough idea of the exposure of individuals moving about in a workroom carrying out a number of operations involving radioactive materials. By proper placement, lengthy observations, occasional use of personal samplers and two-stage samplers, and careful study of past results it is possible for the air sampling program to yield data indicating the probable upper level of exposure of the workers.

It is not as a dose-measuring device that air sampling is most important, but as a tool for locating and limiting air contamination. The same observations listed above as necessary for even a rough exposure estimate, yield far more value in limiting the exposure by better control. Particle size studies may point to a specific operation as the cause of air contamination. Personal air samplers will indicate whether the actions of the worker are the major source of airborne dust. If it is argued that this wide variety of specialized sampling techniques is unnecessary in a particular situation since air contamination is well controlled, then one should ask whether too many samples are already being taken or whether sampling is required at all in that area. Only by making the measurements suggested can one get some idea of the magnitude of the probable error of estimating the average contamination level on a single series of samples or on a group of samples. [22, 27, 28, 29]

Although simple air sampling is not usually an expensive operation, it does cost something. The health physicist should have some idea of the magnitude of these costs. If studies indicate that less sampling is required to achieve a required desirable degree of safety, consideration should be given to whether the money saved could not be used better for health protection in some other area. In other words, the criterion is not the exactness of the dose estimate or the large numbers of samples run or the sophistication of the samplers used, but the degree of health protection provided by the entire health and safety program.

The suggestions here for the use of a wide variety of sampling methods and data study encounter some real practical difficulties. The first one is that concentration levels of concern are so low in some cases that special samplers such as personal samplers or two-stage samplers do not collect enough material to permit accurate analysis. This difficulty is particularly acute in areas where plutonium-239 is handled in enclosed facilities. Long sampling times can be used and numerous samples can be taken even though most of them will yield no information. If no samples containing sufficient material for satisfactory counting are obtained this, too, is important and one should consider whether much or any air monitoring is necessary. In many cases, the only time that high air samples are obtained is when they are expected as a result of maintenance work or transfers of active material. Respiratory protective devices are worn at such times and air samples supply no necessary information.

It is not easy to reduce the size of a sampling program and still ensure its effectiveness and the savings are not very tangible. It is easier to add another sampler or two than it is to decide where to

take one out. We do need to develop some criteria
for the necessity of sampling at a specific location.
For example, is the absence of any sample in excess
of 10 percent of the ICRP permissible level for a year
an acceptable reason for not sampling there?

Lack of adequate health physics manpower is
another practical difficulty. Again, we should look
carefully at our utilization of such manpower. Perhaps, by an intensive study of the air sampling program over a period of a year, the results may permit a drastic reduction in this activity and a net manpower saving. It is not necessary to study all operations at once. One area at a time can be selected for
study. Industrial hygienists and analytical chemists
can give valuable assistance in such studies.

Most of the preceding has dealt with evaluation
of results of the air sampling program. In substance,
they should be used as a source of data for environmental control to reduce air contamination. For
personal dosimetry, their primary aim is to assure
that workers do not inhale quantities of radioactive
materials in excess of accepted ICRP limits. The
precise quantity inhaled, if it is well below these
limits, cannot and need not be measured by air
sampling. One should set an investigation level well
below the ICRP limit and take special note of sample
results in excess of this level. It would seem that
results below this investigation level could simply be
noted as being less than the investigation level and
their exact values not be recorded. Occasional moder-
ate excursions above ICRP values for a single day in
limited working areas certainly should not be regarded as serious incidents requiring lengthy investigations or reports. It is hoped that regulatory

agencies will take a similar attitude. It is possible to create an air monitoring program with the sole object of meeting regulations and still fail to locate sources of serious air contamination.

A growing interest in and a trend toward some standardization in this field should have a favorable influence, provided there is close cooperation among the sponsoring agencies. The report of the ICRP Task Group on General Principles of Monitoring mentioned earlier is an example; another is that of the ICRP Task Group on Lung Dynamics. The U. S. A. Standards Institute has a subcommittee of N-13, which has just completed a state-of-the-art report on air monitoring. This report has been accepted by the full N-13 Committee and will also be submitted to the International Standards Organization. There is an Intersociety Committee on Air Sampling Methods developing standard procedures for use in air pollution studies which has a subcommittee on radioactive substances. The American Public Health Association has a group working on standard methods used in the analysis of radioactive materials in substances of health interest and this group has a subcommittee on air analysis. These last two subcommittees are working together closely and, in effect, functioning as a single committee. The Health Physics Society is officially represented on these.

The European Nuclear Energy Agency included a session on Air Monitoring in its symposium on Radiation Dose Measurements in 1967. The International Atomic Energy Agency sponsored a complete symposium on "Assessment of Airborne Radioactivity" and the proceedings of these symposia are now available. The IAEA also convened a panel of experts to consider this subject and the needs for further study.

One area in which the basic technical aspects of air sampling still present difficult problems is in uranium mining. Better instrumentation is needed to obtain necessary information for the control of radon and its daughter products. Several groups are working on this problem.

It should not be forgotten that air sample data cannot be interpreted in isolation from other pertinent information. Under some circumstances careful surface monitoring may reveal sources of air contamination as quickly as air sampling. Analyses of nose swipes or nose blows are very useful in detecting individual exposures and may also point to failures of respiratory protective equipment. Urine and occasional fecal samples will give the most significant information about individual exposure. Whole body counting is even more useful for gamma emitters. Hopefully, lung counting for plutonium exposure will soon be a practical method to help solve many problems in this field.

I feel that we will know a lot more about mechanisms of air contamination and methods of control when we know more about particle size and activity distributions on airborne material, activity distribution and air concentrations as functions of time and space, and similar data. We will see differences between various types of contaminating processes that will facilitate their control. We may also learn of the limitations of our present control methods.

All of this assumes we have problems in controlling airborne radioactive material. Many problems of this sort are solved by tighter control of environmental conditions without too much regard to identifying the specific sources of the problem. As the

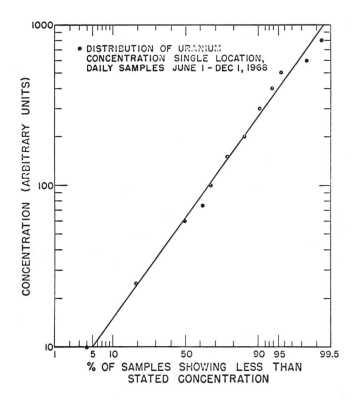

use of radioactive substances spreads to industry, it may become necessary to work closer to established safe limits and costs for all types of health protection, and air pollution control will have to be studied carefully. Air monitoring, then, will play a vital part and we should be prepared to use it wisely.

REFERENCES

1. International Commission on Radiological Protection, Publication No. 12, General Principles of Monitoring for Radiation Protection of Workers, (1969).

2. L. B. Lockhart, R. L. Patterson and W. L. Anderson, Characteristics of Air Filter Media Used for Monitoring Airborne Radioactivity, Naval Research Laboratory Report 6054, (1964).

3. H. J. Ettinger, and J. E. Dummer, Jr., Iodine-131 Sampling with Activated Charcoal and Charcoal Impregnated Filter Paper, LA-3363, (1965).

4. H. J. Ettinger, Iodine Sampling with Silver Nitrate-Impregnated Filter Paper, Health Physics 12, 305-311 (1966).

5. J. J. Fitzgerald, and B. W. Barelli, Determination of Efficiency of the Kanne Chamber for Detection of Radiogases, KAPL-1231, (1954).

6. U. S. Public Health Service Pub. 494, Control of Radon and Daughters in Uranium Mines and Calculations on Biologic Effects (1957).

7. Task Group on Lung Dynamics, Deposition and Retention Models for Internal Dosimetry of the Human Respiratory Tract, Health Physics, 13, 173-207, (1966).

8. P. Kotrappa, Dose to Respiratory Tract from Continuous Inhalation of Radioactive Aerosols, UR-49-964, (1968).

9. P. G. Voilleque, Calculation of Organ and Tissue Burdens and Doses Resulting from an Acute Exposure to a Radioactive Aerosol Using the ICRP Task Group Report on the Human Respiratory Tract, IDO-12067, (1968).

10. B. Shleien, A. G. Friend and H. A. Thomas, A Method for the Estimation of the Respiratory

Deposition of Airborne Materials, Health Physics, 13, 513-517, (1963).

11. B. V. Anderson, Plutonium Aerosol Particle Size Distributions in Room Air, Health Physics, 10, 899-907, (1964).

12. E. C. Hyatt, W. D. Moss and H. F. Schulte, Particle Size Studies on Uranium Aerosols from Machining and Metallurgy Operations, Amer. Ind. Hyg. Assn. Journ. 20, 99-107, (1959).

13. W. D. Moss, E. C. Hyatt, and H. F. Schulte, Particle Size Studies on Plutonium Aerosols, Health Physics, 5, 212-218, (1961).

14. R. J. Sherwood, and D. C. Stevens, Some Observations on the Nature and Particle Size of Airborne Plutonium in the Radiochemical Laboratories, Harwell, Ann. Occ. Hyg. 8, 93-108, (1965).

15. R. A. Kerchner, A Plutonium Particle Size Study in Production Areas at Rocky Flats, Am. Ind. Hyg. Assoc. J., 27, 396-401 (1966).

16. H. E. Ayer, G. W. Sutton and I. H. Davis, Report on Size-Selective Gravimetric Sampling in Foundries, Public Health Service Report TR-39, (1967).

17. G. W. Sutton, and S. J. Reno, Respirable Mass Concentrations Equivalent to Impinger Count Data, Barrè, Vermont, Granite Sheds, Paper presented at Amer. Ind. Hyg. Conf., St. Louis, (1968).

18. H. E. Ayer, The Proposed A.C.G.I.H. Respirable Mass Limit for Quartz: Review and

Evaluation, Trans. Amer. Conf. Government. Ind. Hygienists, (1968).

19. T. T. Mercer, Air Sampling Problems Associated with the Proposed Lung Model, Proceedings 12th Annual Bioassay and Analytical Chemistry Meeting, (1966).

20. H. F. Schulte, Personal Air Sampling and Multiple Stage Sampling; Interpretation of Results from Personal and Static Air Samplers, Proc. Symposium on Radiation Dose Measurements, European Nuclear Energy Agency, Stockholm, (1967).

21. B. Shleien, A Comparison of the Respiratory Deposition of Airborne Activity as Determined by Graded Filtration Technique and the Current ICRP Lung Model, This Conf.

22. R. J. Sherwood, On the Interpretation of Air Sampling for Radioactive Particles, Amer. Ind. Hyg. Assn. Journ. 27, 98-109, (1966).

23. P. N. Dean, and W. H. Langham, Tumorigenicity of Small Highly Radioactive Particles, To be published in Health Physics.

24. R. Caldwell, and T. Potter, The Solubility of Inhaled Particles, Paper presented at the 14th Annual Bioassay and Analytical Chemistry Meeting, New York, (1968).

25. R. J. Sherwood, and D. C. Stevens, A Special Programme of Air Sampling in Selected Areas of A.E.R.E. Harwell, AERE-R-4680, (1964).

26. F. P. J. Robotham, D. C. Stevens, S. F. Pond and I. S. Jones, Release of Plutonium from

Glove Boxes and Fume Hoods, S. R. P. International Symposium on the Radiological Protection of the Worker by the Design and Control of His Environment, Bournemouth, 1966 Paper No. 38.

27. B. A. J. Lister, Development of Air Sampling Technology by the Atomic Energy Research Establishment, Harwell, Proc. Symp. Assessment of Airborne Radioactivity, IAEA, Vienna, (1967).

28. A. J. Breslin, L. Ang, Glauberman, A. C. George and P. Leclare, The Accuracy of Dust Exposure Estimates Obtained from Conventional Air Sampling, Amer. Indust. Hyg. Assn. Journ., 28, 56-61, (1967).

29. A. J. Breslin, Sources of Error in the Methods of Air Monitoring for Inhaled Radioactive Materials, Proc. Symp. on Radiation Dose Measurements, European Nuclear Energy Agency, Stockholm, (1967).

THE MEASUREMENT OF ARGON-41 EFFLUENT FROM UNIVERSITY-SIZE REACTORS

Robert A. Meck
Donner Lab
University of California
Berkeley, California

INTRODUCTION

The MPC for the unrestricted release of argon-41 is $4 \times 10^{-8} \mu Ci/cc$.[1] In practice this concentration is difficult, if not impossible, to measure with commercially available effluent monitors. Adding to this difficulty is the fact that calibration sources of argon-41 are not commercially available. The absolute quantity of ^{41}Ar produced by Oregon State University's TRIGA III reactor was measured by using a sensitive ion chamber and electrometer system. This instrument proved to be a simple and direct approach to argon-41 monitoring. The advantages of this system include, at a relatively low cost, the ability to measure the effluent due to very rapid reactivity pulses and to monitor less than the MPC of

1029

^{41}Ar. This TRIGA is considered representative of university reactors.

EXPERIMENTAL SYSTEM

Schematic diagrams of the experimental system are presented in Fig. 1 and Fig. 2.

At Oregon State University the only significant sources of ^{41}Ar are the four beam tubes, the thermal column, the thermalizing column and the pneumatic transfer tube. The air in other locations is either trapped, in low concentrations or exposed to a low thermal flux. For these reasons, the only measurements of ^{41}Ar were from the air exhausting from these facilities. However, due to the large reactor pool at Berkeley's TRICA III, a significant amount of ^{41}Ar is released from reactor waters. Berkeley's ^{41}Ar release from pool waters at 1 MW approaches $17 \mu Ci/min$. and can not be ignored as in the present case.[2]

The measurement of ^{41}Ar at Oregon State University was made by passing the entire effluent of the argon duct through a Cary Instruments Model 3810 tritium monitor. The exhaust of the rabbit facility was similarly passed through the Cary electrometer system.

Calibration

An argon-41 source was manufactured by irradiating research grade, stable argon in the glory hole facility of the Oregon State University AGN-201 reactor. The thermal neutron flux within the argon

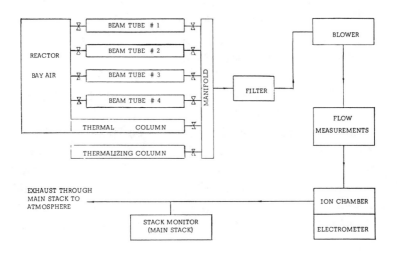

Fig. 1. Schematic: Argon Duct Experimental
Configuration.

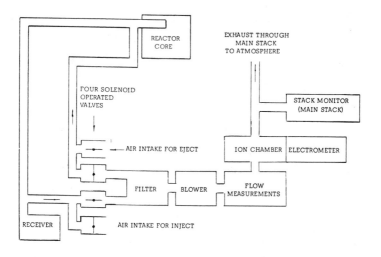

Fig. 2. Schematic: Pneumatic Transfer
(Rabbit) Experimental Configuration.

irradiation container was measured by using NBS calibrated gold foils. The argon irradiation container was simply a polyethylene container with "quick-connect" valves at both ends.

The ^{41}Ar produced was allowed to decay in the ion chamber, and the activity was observed to have the 1.83 hour half-life characteristic of ^{41}Ar. The instrument response was found to be linear on all ranges with a slope of 1.875 x $10^{-8}\mu$Ci ^{41}Ar/cc-mV. This calibration indicated that the MPC of ^{41}Ar would correspond to a reading of 2.13 mV above background. Under stable conditions (i.e. natural radon concentrations relatively invarient) background was measured to be 20 mV within ± 1.0 mV. As shown in Table 1, flow rates through the ion chamber varied from 7.7 cfm to 55.6 cfm depending upon the reactor configuration.

Steady State ^{41}Ar Production Rates

To determine the steady state production rate, the reactor was brought to the desired power in a square wave. Figure 3 illustrates that gamma ray spectra of the effluent are characteristic of argon-41.

It was necessary for the reactor to remain at the same power for several hours in order to approach the ^{41}Ar equilibrium production rate of the argon duct effluent. The concentration, N, of ^{41}Ar in the argon duct was observed to vary with time as given by:

$$N = N_E \ (1 - e^{-\lambda t}),$$

where λ is the effective production constant.

It is possible to calculate the equilibrium value by choosing two observed values. The argon duct

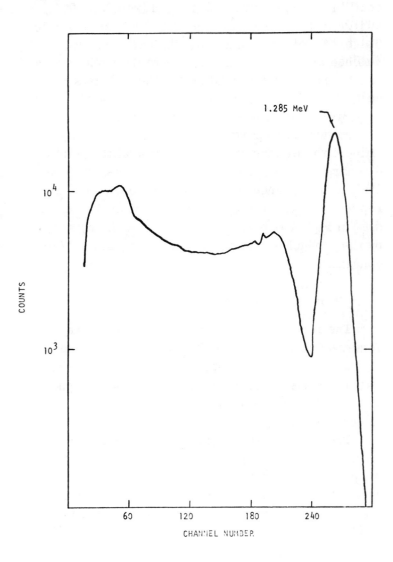

Fig. 3. Gamma Ray Spectrum of Argon Duct
Effluent at 250 KW Reactor Power.

equilibrium concentrations ranged from 3 x 10^{-7} $\mu Ci/cc$ to 1.06 x 10^{-4} $\mu Ci/cc$ at 250 KW as shown in Table 2. Rabbit exhaust equilibrium values were reached almost immediately due to the rapid turnover of the small volume irradiated. Table 3 lists the concentrations recorded from the rabbit exhaust.

Breslin and Dunster, in separate articles, have argued that the reporting of the total radioactive effluent rather than its concentration at the point of discharge is a more realistic basis of assessment.[3, 4] They urge a change in regulations allowing this type of reporting of discharge. In support of these ideas the data in Fig. 4 is presented in μCi $^{41}Ar/min$, as given by the product of the gross flow rates and the ^{41}Ar concentrations.

^{41}Ar Production by Pulses

The resolution of the pulse-produced ^{41}Ar was generally good as seen in Fig. 5 and Fig. 6. In pulse B, the activity was swept from the chamber before it was all counted. Theoretical considerations lead one to conclude that the total argon-41 produced in each case is the same.

The total activity produced by a pulse is a linear function of the energy released as illustrated in Fig. 7. All pulses were initiated from a reactor power of 1 watt. No explanation is offered for the apparent anomoly of ^{41}Ar production in the argon duct for a $2.50 pulse.

TABLE 1.

FLOW RATES

FACILITY	CONFIGURATION	NET FLOW RATE
Argon Duct	Manifold - all open Beam doors - closed Beam valves - closed	7.7 cfm
Argon Duct	Manifold - all open Beam doors - open Beam valves - open	12.3 cfm
Rabbit	Blower on	55.6 cfm
Rabbit	Blower off	7.9 cfm

TABLE 2.

SATURATION ^{41}Ar CONCENTRATION IN THE ARGON DUCT

Reactor Power	$T_{1/2}$ (Production)	Calculated Concentration (uCi/cc)	Extrapolated Concentration (uCi/cc)	Relative Error
1 kW*	41.26 min.	3.23×10^{-7}	3.08×10^{-7}	4.6%
10 kW*	44.72 min.	4.67×10^{-6}	4.67×10^{-6}	0.0%
100 kW*	35.13 min.	6.23×10^{-5}	5.96×10^{-5}	4.3%
250 kW*	56.82 min.	1.58×10^{-4}	1.58×10^{-4}	0.0%
250 kW**	58.74 min.	1.06×10^{-4}	1.06×10^{-4}	0.0%

* Beam doors and valves closed.

** Beam doors and valves open.

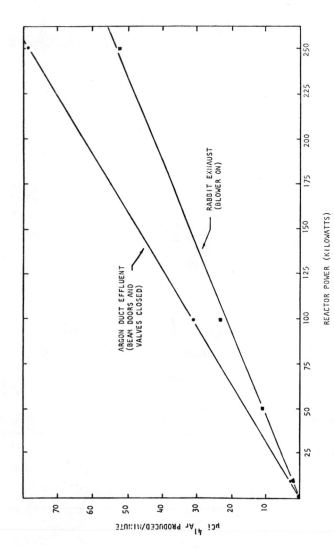

Fig. 4. Production of ^{41}Ar at Equilibrium.

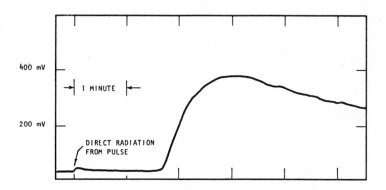

Fig. 5.   Argon Duct Effluent Activity Following
A $3.00 Pulse.

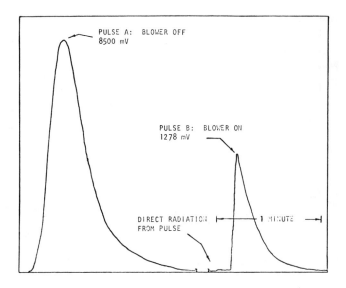

Fig. 6.   Comparison: Rabbit Exhaust Activity
Following A $3.00 Pulse.

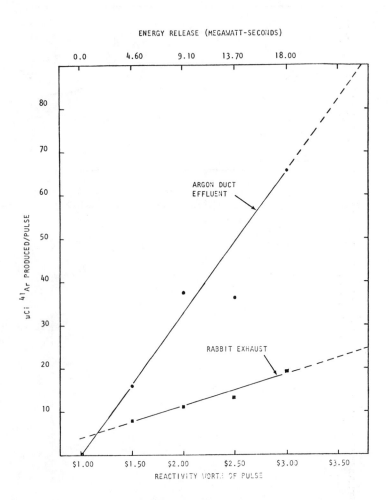

Fig. 7.   Production of Argon-41 Due to Pulses.

DISCUSSION

A maximum of experimental flexibility is desirable with a university reactor. Also, some of the anxieties of residents near the reactor depend upon the reported effluent. If a facility has to report a high concentration of effluent because of insensitive monitoring, some flexibility is lost and public relations suffer.

The sentivities of the monitors listed in Table 4 are somewhat deceptive. For example, the University of California's monitor should detect the argon-41 MPC at 17 cpm above background. This facility presently reports a concentration of $2.4 \times 10^{-7} \mu Ci/cc$ whenever the reactor is operating; i.e. 100 cpm above background. This is due to the width of the recorder trace and unavoidable inaccuracies of a logarithmic circuitry and readout. Oregon State University experiences similar difficulties. The limitations of the logarithmic readout are being circumvented by the installation of a digital system at the Berkeley reactor.

The greater sensitivity of the electrometer-ion chamber system near the MPC is attributed to a slightly more uniform background due to a large sensitive volume and a stable linear readout on a low range. The components for this type of system are available at a slightly lower cost than monitors expressly built for gaseous effluents. The large ion chamber and the option of rapid chart speeds were essential to the measurement of activity produced by rapid reactivity pulses.

Most effluents are monitored by sampling from a large air flow, since it is not practical to monitor the whole exhaust. A large volume ion chamber

TABLE 3.

SATURATION ^{41}Ar CONCENTRATION
IN THE RABBIT EXHAUST

Reactor Power	Blower Condition	Concentration (µCi/cc)
1 kW	ON	Not detectable
10 kW	ON	1.35×10^{-6}
50 kW	ON	6.98×10^{-6}
100 kW	ON	1.48×10^{-5}
250 kW	ON	3.34×10^{-5}
250 kW	OFF	4.07×10^{-5}

TABLE 4.

A COMPARISON OF ARGON MONITORS

Location	Detector	Sensitivity (µCi ^{41}Ar/cc-cpm)	^{41}Ar Equilibrium Production at 250 kW
Cornell Univ. (5)	G-M	2.1×10^{-9}*	----
Oregon St. Univ. (experimental)	ion chamber	1.875×10^{-8}µCi/cc-mV (20 mV bkgrd)	88 µCi/min.
Oregon St. Univ. (6) (stack monitor)	G-M	1.8×10^{-8} (20 cpm bkgd)	68 µCi/min.
Univ. of Cal. Berkeley (7)	NaI	2.4×10^{-9}* (100 cpm bkgd)	204 µCi/min.*
Univ. of Ill. (8)	G-M	1.2×10^{-9} (100 cpm bkgd)	13.6 µCi/min**
Wash. St. Univ. (9)	G-M	4×10^{-9}* (67 cpm bkgd)	85 µCi/min.*

* Calculated from reported data.

** Of the reactor facilities listed, the one at the Univ. of Ill. is most similar to that of Oregon State University. The volume of effluent is 1/5 that of Oregon State University's, while the ^{41}Ar concentration released to the atmosphere is reportedly the same at both facilities.

enables one to sample a larger volume than usual and at atmospheric pressure. At Berkeley the effect of a large volume is accomplished by monitoring a sample at 10 atmospheres; however, experience has proved that the compressor has to be maintenanced every few months.

While the calibration of monitors for argon-41 is best done with the identical isotope, potassium-40 has similar emissions and a long half-life. For routine routine calibrations a potassium-40 source can be used.

Experience at the University of California and Oregon State University has shown that the production of argon-41 can be minimized by tightly sealing openings and holes in facilities such as the thermal column and medical facilities.

CONCLUSION

For university-size reactors, a sensitive electrometer ion-chamber system can be employed for gaseous effluent monitoring of argon-41 at or near MPC and during rapid reactivity pulses.

ACKNOWLEDGEMENT

The author is grateful for the reactor time and assistance allotted by Drs. C. H. Wang and J. C. Ringle, Radiation Center, Oregon State University, Corvallis, Oregon.

BIBLIOGRAPHY

1. National Committee on Radiation Protection. Maximum Permissible Body Burdens and Maximum Permissible Concentrations of Radionuclides in Air and in Water for Occupational Exposure. Washington, D. C. U. S. Government Printing Office, 1959. 95p. (U. S. National Bureau of Standards Handbook 69).

2. Lee Stollar, Reactor Supervisor, University of California, Berkeley, TRIGA III Reactor; Personal Communication; Berkeley, California, January 3, (1969).

3. A. J. Breslin, Health Physics 12 : 1496 (1966).

4. H. J. Dunster, Health Physics 13 : 916 (1967).

5. (anonomous), Nucleonics 21, (2) : 48 (1963).

6. John C. Ringle, Assistant Reactor Administrator, Radiation Center, Oregon State University; Personal Communication; Corvallis, Oregon, August 18, (1967).

7. Lewis Hughes, Radiological Safety Officer, University of California, Berkeley; Personal Communication; Berkeley, California, August 18, (1967).

8. Paul R. Hesselman, Health Physicist, Nuclear Engineering Laboratory, University of Illinois; Personal Communication; Urbana, Illinois, September 27, (1966).

9. Roger C. Brown, Health Physics 9 : 315 (1963).

ARGON-41 MEASUREMENTS AT A
POOL REACTOR

John D. Jones
Michigan Memorial-Phoenix Project
Phoenix Memorial Laboratory
The University of Michigan

ABSTRACT

An accurate, inexpensive, continuous air monitoring device for measuring argon-41 and other radioactive noble gases has been constructed from a 55-
gallon drum and a Geiger-Müller counter. Argon-41
concentrations in the building air and the exhaust stack
effluent are of primary interest. Radon progeny and
other particulate contaminates are filtered by a membrane filter before the air enters the drum. Background variations are minimal. Simple calibration
procedures are outlined. The sensitivity of the
monitor is sufficient to determine concentrations
less than 2% of the occupational MPC_a for ^{41}Ar.

^{41}Ar activity increases in the reactor building
were correlated with the bulk pool water temperature.
The occasional absence of detectable ^{41}Ar activity was

attributable to thermal layers in the pool. The effect of a calorimetric calibration of the reactor on argon released is described. ^{41}Ar concentrations in the discharge stacks were correlated with reactor core configuration near the pneumatic sample tube system. Four years experience with the device at a 2 megawatt pool-type research reactor are summarized.

INTRODUCTION

A study of the ^{41}Ar produced and released by the 2 Mw Michigan Memorial-Phoenix Project's Ford Nuclear Reactor (FNR) at The University of Michigan was initiated in early 1965.

The purpose of this study was:

1) to develop and calibrate a monitor which could be used to continuously measure with accuracy and dependability, concentrations of ^{41}Ar as low as the non-occupational MPC_a;

2) to determine the average ^{41}Ar concentration produced by the reactor and released from the building exhaust stacks;

3) to determine the average concentrations in the building air and study the means of release of argon activity from the surface of the reactor pool.

^{41}Ar, having a half-life of 108 minutes, is produced by thermal neutron capture in ^{40}Ar, whose natural isotopic abundance is 99.6%. The cross section is 0.54 barns. ^{41}Ar undergoes coincident beta and gamma decay. The maximum beta ray energy is 1.25 MeVand the gamma energy is 1.3 MeV. Air contains 0.934 mole percent argon. Argon is chemically inert.

CONSTRUCTION OF THE MONITOR

The dependability and high sensitivity of a Geiger-Müller counter for beta rays, as energetic as those from ^{41}Ar, make the GM tube a logical choice as a detector. The maximum range of the 1.25 mev beta particle in air is 170 inches. Providing the detector with a large volume of the gas thus increases the sensitivity. A 55-gallon drum with a side bung was selected because it was a convenient and inexpensive device in which a GM tube could be installed. (Fig. 1) The gamma rays from the ^{41}Ar are of little significance since the GM tube has a very low efficiency for detecting gamma rays of this energy. The GM tube selected, primarily because of its low cost and thin aluminum walls, was a Victoreen type 1B85. The drum used was an ICC17E drum with a volume of 58 gallons (2.2×10^5 cc).

A large rubber stopper plugs the top bung and a copper tube which reaches to the bottom of the drum serves as the inlet. A short copper tube inserted in the rubber stopper serves as the exhaust tube.

To prevent contamination of the drum and to catch dust laden with radon and thoron progeny, a Gelman type GM-4 membrane filter is attached to the intake side of a small carbon vane pump used to continuously

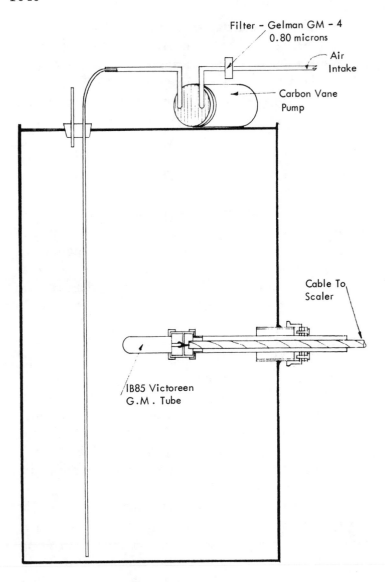

Fig. 1. Sectional view of monitor; air is continuously drawn through a membrane filter to eliminate radon and thoron progeny as it is pumped through the drum.

pump air through the drum. The filter has a pore
size of 0.80 microns. It was found to be effective in
eliminating background variations due to natural
radioactivity. The flow rate through the drum is
approximately 0.54 ft^3/min. The average hold up
time in the drum is thus about 14 minutes.

The GM tube is connected to a conventional scaler
ratemeter assembly. Digital data from the scaler is
used to compute average concentrations of ^{41}Ar over
extended periods of time. The ratemeter and recorder
provide instantaneous information on gaseous activity
levels.

CALIBRATION OF THE SYSTEM FOR ARGON-41

Gamma Spectrometry

The first calibration of the monitor was done by
independently determining the gamma activity of a
representative sample of the gas from the drum. A
glass cylinder 2" x 3-3/4" I.D., having side arms
1/4" I.D. and a lucite window 1/16", was connected
in series with the drum by means of a short rubber
tubing. Exhaust air from the reactor pneumatic tube
system, provided a convenient source of ^{41}Ar. An
estimated concentration of approximately 1.5 x 10^{-5}
μCi/cc, was passed through the monitor drum and
cylinder for several hours to insure equilibrium between the drum and cylinder. A count was then taken
on the monitor scaler and the cylinder removed quickly
to a shielded NaI detector for calibration. (Fig. 2)

^{41}Ar activity in the cylinder was determined by
placing the center of the cylinder at a distance of 10 cm

JOHN D. JONES

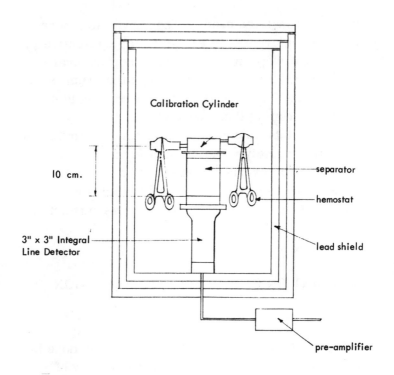

Fig. 2. Calibration assembly for independently determining the gamma activity of a representative sample of gas from the drum.

from the face of a Harshaw 3" x 3" NaI integral line detector. Analysis was done using a 400 channel analyzer. A photopeak efficiency K of 5.61 x 10^{-3} was derived from Heath.[1]

$$\mu\text{Ci/cc (cylinder)} = \frac{\text{cpm photopeak (A)}}{\text{cylinder vol. (339.1 cc) x K x 2.22 x } 10^{6} \text{ dpm}/\mu\text{Ci}}$$

The calibration factor to be used when the monitor is in use is:

$$\mu Ci/cc - cpm = \frac{\mu Ci/cc \text{ (cylinder)}}{\text{Monitor cpm (B) x D x H}}$$

B = cpm monitor

D = decay correction (removal from drum to time of calibration)

$$H = I - e^{-\lambda t}$$

where

$$t = \frac{\text{Drum vol. (cc)}}{\text{Flow rate cc/min.}}$$

Results are shown in Table No. 1.

TABLE 1

Calibration of Monitor Using Gamma Spectrometry Method

Total Counts Photo-Peak	Time (min.)	A	H	B	$\mu Ci/cc - cpm \pm$ % Std. Dev.*
1245	15	83	.92	7839	$2.90 \times 10^{-9} \pm 3.75\%$
973	15	64.7	.92	7022	$2.52 \times 10^{-9} \pm 4.1\%$
1035	15	69	.92	6982	$2.71 \times 10^{-9} \pm 4.0\%$
				Average	2.71×10^{-9}

*Error estimates based on counting statistics

Proportional Counter Method

A method similar to that of Day and Selden[2] was also used to calibrate the drum. A sealed 2-inch diameter 2π proportional counter (Nuclear Measurements Corporation) was set up as shown in Fig. 3.

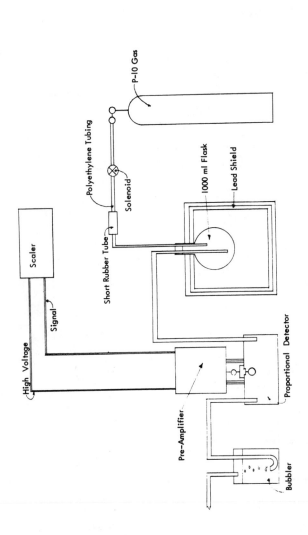

Fig. 3. Schematic diagram showing apparatus for calibration of a small volume of ^{41}Ar in a proportional counter. After calibration has been accomplished, the ^{41}Ar is transferred by suction to the monitor drum.

When a calibration was to be done, the system was purged with P-10 gas* or methane. A small (2 cc) snap cap polyethylene vial was filled with argon and irradiated in the pneumatic rabbit system of the FNR for about one minute producing an estimated 10 μCi of ^{41}Ar. A syringe was used to withdraw a small sample of the gas ($\sim$.5 cc), and this was quickly injected into the rubber tubing. The counter was flushed after injection of the ^{41}Ar until a count rate of approximately 100,000 cpm was obtained. At this time, the hemostats were used to clamp the short rubber hoses near the detector so that there was no further change in the counter's contents. After a plateau was made of the proportional counter using the ^{41}Ar as the radiation source, the activity of the sample was determined and then transferred into the drum by suction. Experience showed that the beginning and ending of the plateau varied slightly with each new calibration, probably due to small amounts of air injected into the system. It is assumed that the proportional counter detects 100% of the beta activity.[2] In the drum, homogeneity is, for all practical purposes, established within 5 minutes. This was confirmed by observing that after 5 minutes the activity in the drum followed the decay curve for ^{41}Ar. Counts for the drum calibration were thus started from 8 to 10 minutes after the sample was transferred to the drum.

The same method was used with methane as a counting gas. The data (Table 2) indicate no appreciable difference due to using the two gases.

Summary of Calibration Methods

When a multichannel analyzer is available, the in-line scintillation method appears to be slightly

---

* 90% argon, 10% methane

TABLE 2

Calibration of Monitor for Argon-41 Using Proportional Counter Technique

Sample	Gas	High Voltage	Proportional Counter – cpm	Net Monitor – cpm Corrected to 0 Time	μCi/cc – cpm ± % Standard Deviation*
1	P-10	2000	225,504	160	$3.17 \times 10^{-9} \pm 3.6\%$
2	P-10	2000	90,428	88.6	$2.40 \times 10^{-9} \pm 4.8\%$
3	P-10	1900	89,924	73.6	$2.72 \times 10^{-9} \pm 5.2\%$
4	CH_4	2900	98,438	95	$2.31 \times 10^{-9} \pm 4.6\%$
5	CH_4	2900	43,818	41.3	$2.37 \times 10^{-9} \pm 7.0\%$
6	CH_4	2800	101,232	105	$2.39 \times 10^{-9} \pm 4.4\%$
7	CH_4	3500	217,037	146	$3.30 \times 10^{-9} \pm 3.7\%$

Average 2.66×10^{-9}

* Errors based on counting statistics

easier to perform with a minimum of disturbing influences. The method of transferring a sample of gas previously calibrated by the proportional counter must be done with extreme care to avoid sample transfer problems which could lead to erroneous answers. Additionally, there is greater inconvenience in setting up the required equipment.

BACKGROUND AND G. M. TUBE INTERCHANGEABILITY

The drum can be placed in any convenient low background area. Since a long hose can be connected to the drum and air sampled in areas quite remote from the actual detector, lack of mobility of the drum poses little problem. Background measurements over a 4-year period varied from 49 cpm to 61 cpm.

Seven 1B85 CM tubes were tested for interchangeability. As air from an exhaust duct was sampled by the monitor, each tube was mounted in the drum and the average of two, 2-minute counts were taken. As shown in Table 3, the greatest deviation from the mean was only 5.3% and the standard deviation from the residuals was only 3.1%.

MONITORING OF PNEUMATIC TUBE EXHAUST SYSTEM AT FNR REACTOR SITE

At any reactor where air is used to operate a pneumatic rabbit system, significant amounts of ^{41}Ar are produced. Figure 4 shows how the pneumatic tubes are connected to the building exhaust stack.

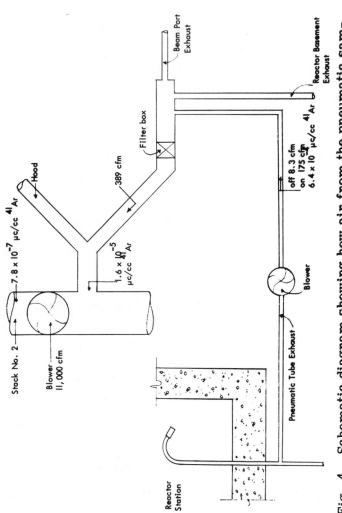

Fig. 4. Schematic diagram showing how air from the pneumatic sample tube system is connected to the building exhaust system.

The building exhaust blower exhausts air at 11,000 cfm and exerts a negative pressure on the pneumatic tube system causing air to flow through the tubes at 8.3 cfm.   This flow is sufficient to produce a constant ^{41}Ar concentration in the exhaust stack whether or not the pneumatic blower is on.

The monitor was used to measure the average ^{41}Ar concentration released from the Phoenix Building exhaust stack by sampling the stream of air just before it is exhausted.   These measurements were started in January of 1965 and have been continued to the present.

Figure 5 summarizes this data.

During normal 2 Mw operation, the argon concentration released from the exhaust stack is relatively constant.   The beam tubes, which are vented into the same exhaust system,   could also contribute to the ^{41}Ar released; however, flow of air through these tubes is normally insignificant.   In addition, each beam tube is usually backfilled with He or N_2.

It has been found that different core configurations change the ^{41}Ar production rate in the pneumatic tube system.   Before a D_2O tank was installed at the rear of the reactor core in February 1965, the average ^{41}Ar concentration in the exhaust stack at 2 Mw, was 7.4×10^{-7} μCi/cc.   After the D_2O tank was installed, the concentration increased to 9×10^{-7} μCi/cc in April of 1965.   This increase was primarily due to the placement of the core resulting in a different flux distribution (see Fig. 6).

In may of 1965, a new configuration of the core (Fig. 7) which substituted fuel elements for two reflectors which were between the core and the pneumatic bundle, resulted in an activity increase in the exhaust

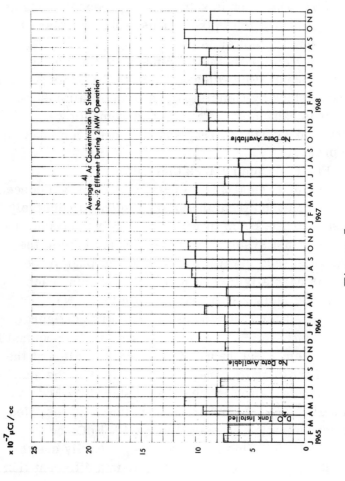

Figure 5.

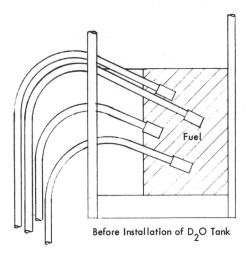

Before Installation of D_2O Tank

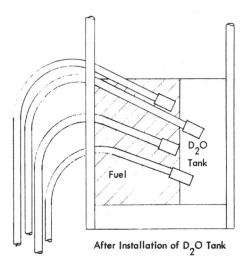

After Installation of D_2O Tank

Fig. 6.   FNR core before and after D_2O tank was installed in February 1965. [41]Ar exhausted from pneumatic system increased by approximately 30% due to different flux distribution.

JOHN D. JONES

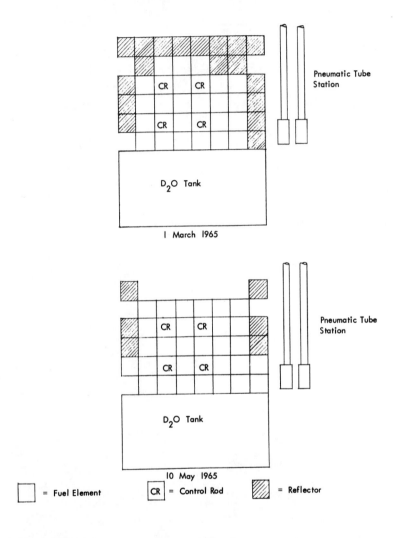

Fig. 7. Reactor core configuration relative to pneumatic tube station. [41]Ar concentration in exhaust system increased by approximately 8% in May 1965, due to the new flux distribution.

system to 1.10×10^{-6} μCi/cc. The monitor easily detected these changes in argon concentration. Operational procedures such as closing off one or more pneumatic tubes for repair were responsible for reducing the ^{41}Ar concentration during some months.

ARGON ACTIVITY IN REACTOR POOL DURING CONTINUOUS 2 MW OPERATION

The solubility of argon in water at various temperatures is shown in Fig. 8. It is 2-1/2 times as soluble as nitrogen and has about the same solubility as oxygen.[3] Within the normal temperature range of the FNR pool (92° - 108°F.) the solubility varies by about 9%.

Direct measurements of ^{41}Ar in samples of water from the pool were made by bubbling argon through a 100 ml sample of freshly collected pool water into the calibrated drum. Results are shown in Table 4.

Assuming a uniform distribution of ^{41}Ar throughout the pool, the total argon activity in the pool is approximately 234 millicuries.

An interesting condition occurs when the temperature of the water returned to the pool (Fig. 9) is less than the temperature of the pool at the surface. Under these conditions, the argon is trapped in a thermal layer and the activity of argon at the surface, as shown in Table 4, is less than 3×10^{-5} μCi/ml. This results in a negligible ^{41}Ar concentration in the room air.

JOHN D. JONES

TABLE 3

Interchangeability of G. M. Tubes

Tube	Counts/2 Min.	$x - \bar{x}$	$(x - \bar{x})^2$
1	2119	−118	13,924
2	2208	−39	1,521
3	2260	+23	529
4	2316	+79	6,241
5	2183	−44	1,936
6	2282	+45	2,025
7	2288	+51	2,601
$\bar{x}$	2237		28,777

$$\sigma \bar{x} = \pm 47.2 \ (2.1\%)$$

σ from residuals $\pm 69.2 \ (3.1\%)$

TABLE 4

Measured ^{41}Ar Activity in FNR Pool During 2 MW Operation

Primary Heat Exchanger Exit Temperature ($^{\circ}$F.)	2 Ft. Below Pool Surface Temperature ($^{\circ}$F.)	Δt	Sample Location	μCi/ml. (H_2O)
102.6	101.5	(+)	Pool Surface	1.3×10^{-3}
104	103.7	(+)	Pool Surface	1.27×10^{-3}
104.2	105.8	(−)	Pool Surface	$< 3 \times 10^{-5}$
102.2	104.8	(−)	Pool Surface	$< 3 \times 10^{-5}$
101.2	104	(−)	Hot D. I. Return	1.59×10^{-3}
100.9	103.9	(−)	Pool Surface	$< 6 \times 10^{-5}$
101.2	100.2	(+)	Pool Surface	1.13×10^{-3}

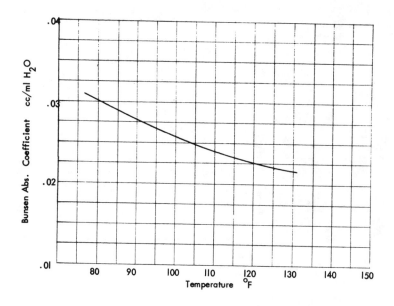

Fig. 8. Solubility of Argon in Water. The
Normal Temperature Range of the FNR Pool
is 92° F. - 108° F.

ARGON ACTIVITY IN REACTOR BUILDING
DURING 2 MW OPERATION

Measurements of the argon concentrations in the
exhaust air from the Reactor Building (Fig. 10) were
done by using the monitor. Because of diluting air
from the rest of the building, [41]Ar activity on the
pool floor is actually 5.5 times higher than the values
measured in the exhaust air. These measurements,
taken over long periods of time (from 200 to 10,000
minutes) confirmed the direct relationship between the
bulk pool temperature and the argon activity in the air.

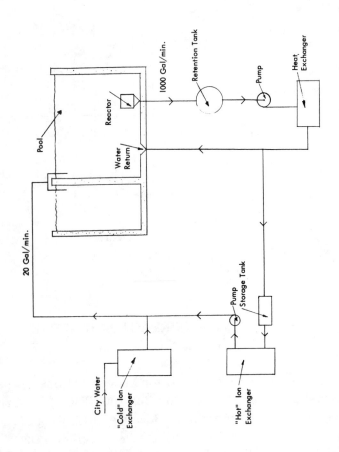

Fig. 9.   General Arrangement of Primary Water System FNR.

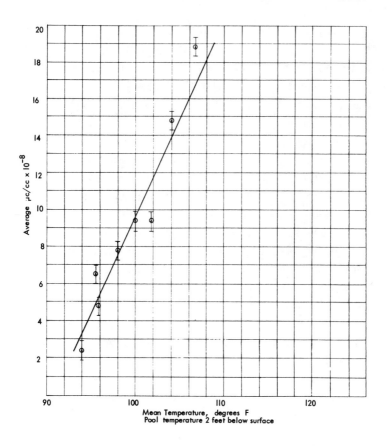

Fig. 10. [41]Ar concentration measured in Reactor Building exhaust system at various pool temperatures. Because of diluting air from the rest of the building, pool floor concentrations are 5.5 times higher than the values presented.

The average argon activity in the air appears to be directly proportional to the average pool temperature; however, minor temperature increases seem to cause rather large "pulses" of argon to be released, due probably to the reduced solubility at higher temperatures.

An example of this phenomenon is given in Fig. 11, which is a plot of the ^{41}Ar concentration in the pool floor area during a 75-minute calorimetric calibration of reactor power. For calibration, secondary cooling is turned off and the temperature of the pool gradually increases from 88.5°F. to 110.5°F. until secondary cooling is restored. The ^{41}Ar concentration in the room air rises until after cooling is restored, whereupon the activity drops rapidly from a peak concentration of 2.67×10^{-6} μCi/cc as the cool water returned to the pool establishes a thermal layer in the pool. The room exhaust system sweeps the argon activity from the area at a rate of 1,680 cfm.

CONCLUSIONS

^{41}Ar concentrations in the exhaust and ventilation systems of a building housing a 2 MW pool reactor were determined. The monitor consisted of a Geiger-Müller detector mounted in a drum. The system described has a sensitivity sufficient to detect 4×10^{-8} μCi/cc ^{41}Ar with a net count rate of 15 cpm.

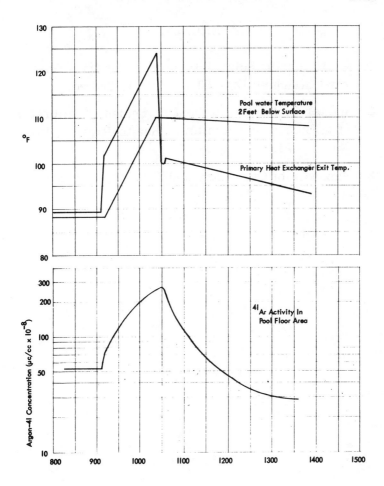

Fig. 11. ^{41}Ar concentration in pool floor area during a 75 minute calorimetric calibration of reactor power.

REFERENCES

1. R. L. Heath, Scintillation Spectrometry - Gamma ray Catalogue IDO-16880, Vol. 1, 2nd ed.

2. Walter C. Day and Robert F. Selden, "Calibration of a Fission Gas Monitor," Health Physics, Vol. 9, No. 9, 1963.

3. Handbook of Chemistry and Physics, 26th ed.

MONITORING OF GASEOUS EFFLUENTS
FROM REACTORS

T. F. Johns
United Kingdom Atomic Energy Authority
Great Britain

INTRODUCTION

The radionuclides most commonly encountered
in substantial quantities in gaseous effluents from
reactors are those produced by activation of mater-
ials in or near the reactor core, e.g. ^{41}A, ^{13}N, ^{19}O;
most of these are either not taken up by the body to
any appreciable extent, or (if they are taken up by the
body) have half lives which are so short that the organ
doses resulting from the uptake are small. People
living or working near a reactor discharging such
gases are irradiated mainly as a result of the fact
that they are periodically in (or under) the plume of
radioactive gas being discharged; i.e. they are exposed
to external irradiation from these materials. The
gaseous effluent may also contain the noble fission
gases Krypton and Xenon, where again the hazard is
one of external irradiation from a cloud of radioactive

gas. Releases of these gases* must be monitored to ensure that the external radiation doses to people living near the reactor are acceptably small. Some characteristics of the gases commonly encountered are shown in Table I.

The gaseous effluent may also contain ^{131}I, and if cows are grazed near the reactor, the release of this nuclide will have to be measured and controlled to avoid excessive doses to the public from drinking contaminated milk. In many cases, the acceptable release rate of ^{131}I will be several orders of magnitude less than the permissible release rate of inert gases.

Techniques developed and used at the Atomic Energy Establishment, Winfrith, for monitoring discharges of inert gases from reactor systems are described, as well as techniques for continuously monitoring for small (millicurie) releases of ^{131}I in the presence of much larger quantities, (hundreds of curies) of inert gases. Difficulties experienced in the use of these monitors are also discussed, and brief mention is made of the technique used for measuring releases of tritiated water vapour.

---

*For the sake of brevity these gases are referred to below as "inert" gases. Strictly some of them (e.g. ^{19}O) are not inert, but in all cases exposure is due primarily to external radiation from a cloud of gas.

Table I

Main Constituents of Gaseous Effluents

Nuclide	Half-Life	Decay Mode	Prominant Gamma Ray energy MeV	Beta Ray Energy (max) MeV
^{16}N	7.4 sec.	β^-	6-7	4 (etc)
^{19}O	29 sec.	β^-	0.2, 1.4	3-4
^{13}N	10 min	β^+	(0.51)	1.2
^{41}A	110 min	β^-	1.3	1.2
^{135m}Xe	15 min	I.T.	0.53	-
^{87}Kr	78 min	β^-	0.4	3.8
^{88}Kr	2.8 hrs	β^-	0.19	0.5
^{85m}Kr	4.4 hrs	β^-	0.15	0.8
^{135}Xe	9.2 hrs	β^-	0.25	0.9
^{133}Xe	5.3 days	β^-	0.081	0.35

RELATIONSHIPS BETWEEN DISCHARGE RATE
AND PERSONAL EXPOSURE

The purpose of stack monitoring is to establish that the doses received by the public, resulting from the discharge, are acceptably small. At Winfrith we have derived working limits for the continuous discharge rates of the more important nuclides, i.e. the discharge rates which, if continuous, would result in doses equal to the I.C.R.P. dose limits being received by the few people living close to the site. The calculation takes account of variability of wind direction and atmospheric stability; but of course such calculations are imprecise and we have deliberately made some conservative assumptions, which result in a considerable over-estimate of the resulting doses. Because we have made these conservative assumptions, we are prepared to operate with discharges as high as the calculated permissible rates, without introducing additional arbitrary safety factors.

In practice the discharge rates are variable, and consequently it is necessary to specify not only the average discharge rate, but also to limit in some way the amount which can be released in a period during which the wind direction might remain unchanged; otherwise there would be a danger that a person downwind during a short period when the release rate was grossly abnormal might receive an unacceptably large dose during that short period. We have taken care of this by giving a working limit for the average rate of discharge and also specifying an accpetable averaging period (never more than 4 weeks and in general one week).

If cows are grazed fairly close to reactor sites, ^{131}I discharges may have to be limited to about 10-20 mCi/day.[1] Discharge rates for inert gases can be much higher, but may have to be limited to the order of 1000 Ci/day if people live within 500 m of the reactor site and the material is released near to ground level. Releases can, of course, be much greater if the gases are released from a tall stack.

Typical values for the acceptable average discharge rates of the more important nuclides from a reactor site with a short stack and people living within 500 m are shown in Table II.

Table II

Typical Derived Working Limits For Gaseous Discharge - Winfrith Site

Nuclide	Acceptable Average Discharge Rate
Inert Gases	1000 Curie MeV/day
[131]I	10 mCi/day
[3]H (as water)	100 Ci/day

The dose received by a person immersed in a large cloud of inert gas is proportional to the product of the concentration c_N of the gas in the air, multiplied by the energy E_N absorbed in the body per disintegration. E_N is what I. C. R. P. call the effective energy per disintegration. It follows that the dose resulting from the cloud of gas containing a number of

radionuclides is equal to the sum over all the gases
present of the product $c_n E_n$, and hence to

$$\sum c_N E_N$$

where c_N is the number of curies released per unit
time. The ideal stack monitor should therefore give
a reading proportional to $\sum c_N E_N$; but there is no
simple way of measuring the total curie MeV content
of a mixture of beta emitters. We have generally
determined the gamma and beta components separately.
The gamma component is straightforwardly determined
by measuring the dose rate at a known distance from
a fixed volume of the gas being discharged; the beta
component has been measured by passing some of the
gas through a beta chamber to measure the concentration and multiplying by an assumed mean beta energy.
Experience has shown that the beta and gamma components of the exposure are in general about equal.

Although, in calculating the maximum permissible
concentrations in air of the noble gases, I.C.R.P.
have regarded the irradiation by the beta component
as essentially a whole-body exposure, and hence given
a precedent for following the above procedure, in
fact the beta component only irradiates the skin and
superficial tissues. Since the permissible dose to the
skin is 6 times the permissible dose to the whole body,
then (if the gamma and beta components of the dose
are about equal) it is clear that if the gamma dose
alone is controlled to ensure that whole body dose
limits are not exceeded, it will automatically follow
that the beta plus gamma dose to the skin will in turn
be well below the corresponding dose limit for the
skin. For this reason we feel that a convincing case
can be made for suggesting that in future one should

only monitor the total gamma curie MeV content of the discharge, (together perhaps with an occasional check of the beta/gamma ratio). This would lead to a considerable simplication in the equipment used, and also to a simplification in the interpretation of the monitor readings.

MONITORING FOR I 131

Millicurie quantities of ^{131}I can be measured, even when accompanied by hundreds or thousands of curies of inert gases, by passing a known fraction of the gas being discharged through a sampling pack containing activated charcoal and designed to absorb the iodine; little absorption of inert gases will occur. The amount of ^{131}I collected on the pack is continuously monitored using a NaI (T1) scintillation counter. The success of the technique depends on the fact that the scintillation detector is detecting gamma rays from all of the iodine collected on the pack since (say) the beginning of the day, whereas of course it only sees radiation from the small amount of inert gases absorbed on or passing through the charcoal at any one time. To achieve some further discrimination against the small amount of inert gas passing through the charcoal or absorbed on it, the NaI crystal is set up to count only gamma rays in a channel centered around 364 keV.

MONITORING FOR INERT GASES

In the Dragon High Temperature Gas Cooled Reactor we have used (sealed) ionisation chambers

placed inside a large volume through which the effluent
passes to measure the gamma component of the inert
gas discharge, and passed some of the gas through
another ionisation chamber to measure the beta component.

In the S.G.H.W.R.* discharges to atmosphere
arise almost entirely from gases passing down the
condenser off-gas line. Activation gases produced in
the reactor core, together with noble fission gases
from any faulty fuel present, pass in the steam directly
to the turbine and its associated condenser; a substantial fraction of these gases is then discharged to atmosphere in the small volume of uncondensed gas
passing from the condenser. A delay is deliberately
introduced between the condenser and the discharge
point to allow the shorter-lived gases (mainly ^{16}N)
to decay before discharge. The gamma component of
the discharge is measured by using a NaI crystal to
detect the gamma rays from a small volume through
which a known fraction of the gas is passed at a known
rate. Energy selection of these gamma rays enables
one to distinguish between high energy radiation from
^{16}N and lower energy radiations from other nuclides.
Only the latter are important in determining the doses
to people off-site, since ^{16}N decays very rapidly in
the first few hundred metres after discharge. In
practice we have found that most of the ^{16}N has de-
cayed before even reaching the stack, and consequently
it has not been of great practical importance to distinguish the high energy radiations due to ^{16}N from
the lower energy gammas from other nuclides. We
have found this monitor very reliable and useful as an

---

*Steam Generating Heavy Water Reactor.

indicator of trends in the discharge rate, but because of the very poor energy response of the NaI detector, there is not a unique relationship between the readings of the instrument and the curie MeV content of the discharge. Consequently we have now fitted an additional monitor. This consists very simply of a gamma monitor, normally used as a portable instrument and incorporating an energy-compensated geiger counter, placed adjacent to the bottom of the stack through which these gases are discharged to atmosphere. Provided that the airflow through the stack is constant, there is a simple linear relationship between the reading of this monitor and the gamma curie MeV release per unit time. The more sophisticated instrument described above continues to be used to give reliable indications of changes in the discharge rate.

MONITORING OF TRITIATED WATER VAPOUR

We also monitor the S. G. H. W. R. stack gases for releases of tritiated water vapour. The technique is similar to that which we are increasingly using for monitoring for tritiated water vapour in air. A known small fraction of the air being discharged is continuously sucked through a bubbler containing water; the tritiated water vapour exchanges with the water and is trapped. [2] Periodically a known volume of the water in the bubbler is mixed with liquid scintillator, and the amount of tritium collected is measured in the usual way.

CALIBRATION OF MONITORS

The iodine monitor is easily calibrated by placing a known amount of iodine on the charcoal.   In
general the sensitivity of inert gas monitors (beta or
gamma) has been determined prior to operation by
the use of ^{85}Kr, a known quantity of this being released into a known volume.

The monitor on the stack of the S. G. H. W. R. has
been calibrated in situ.   During reactor operation, a
sample of the gas being discharged via the condenser
off-gas line was collected in a bottle, and the concentration of each of the gases present was determined
by in situ gamma spectrometry, using NaI and lithium
drifted germanium detectors.   From these measurements it was possible to determine the release rate
of each of the radioactive gases present, and hence
the total release rate in curie MeV/day.   This was
related to the reading on the stack monitor to provide
an appropriate calibration factor.

Calibration of the tritium monitors presents no
difficulties.

EXPERIENCE IN THE USE OF THESE MONITORS

We have had experience of using monitors of the
type described on the Dragon and S. G. H. W. reactors.
In the former case, releases have been very small,
and the monitors have consistently given readings
near to zero.   In the case of S. G. H. W. R. (as with
other direct cycle boiling water reactors) there are
always appreciable releases of activation gases,
including ^{19}O and ^{13}N, and, on occasions when the

reactor has operated for short periods with faulty
fuel, of Kryptons and Xenons.

Normally discharges are substantially less than
the derived working limits, but on one or two occasions when faulty fuel has been present in the reactor
discharges have been near the operating limits; the
discharges have however never exceeded those limits,
when averaged over the appropriate period. The delay
which already exists between the condenser and the
release point is shortly to be increased by modifications to the plant. This will allow some of the shorter-
lived nuclides present to decay further, and hence
lead to a reduction in the amount of radioactive
material released and to increased flexibility in
operating the reactor.

In the S. G. H. W. R. very little iodine escapes from
faulty fuel during operation, and what does leak out
tends to remain in the water phase; very little gets
into the steam phase and still less escapes past the
condenser. Consequently releases to atmosphere
have always been well within the operating limits.

Some difficulty has been experienced because of
the absorption of the short-lived activation products
on the charcoal of the iodine monitors. This leads
to a high "background" reading. We have as yet not
identified the exact chemical and physical form of the
material being absorbed.

Measurements of tritiated water vapour being
released have proved to be quite a sensitive method
of indication of the presence of small leaks of heavy
water; indeed it is true to say that the stack monitoring
system as a whole has been extremely useful not
only in ensuring that discharges to atmosphere are
adequately controlled, but in giving sensitive indication

of the state of the plant. Large sudden increases in the rate of release of inert gases have been shown to correspond to the onset of small defects in fuel pins.

In addition to the continuously operating systems described, there are a number of sampling points at which we have installed sampling packs containing charcoal, which can be removed and counted in the laboratory. These have been useful for the detection of ^{138}Cs and ^{88}Rb, which are short-lived daughters of ^{138}Xe and ^{88}Kr, indicating the presence of fission products.

REFERENCES

1. Pamela M. Bryant, Health Physics, 10, p. 294 (1964)

2. Studies and Techniques in Tritium Health Physics at CRNL, R. V. Osborne; Chalk River Report AECL-26! (November, 1967).

RADIOGAS MONITORING AND
LEAK DETECTION

G. H. Liebler, W. F. Nelson, and J. Tamburri
Health Physics Division, Indian Point Station,
Consolidated Edison Company of N. Y., Inc.

ABSTRACT

The existance of pockets of radioactive contamination within enclosed cells at a nuclear facility may
remain undetected when full reliance is placed on air
monitoring of the ventilating systems. Monitors of
these types can detect concentrations of radiogas as
low as 10^{-7} μCi/cc in the main branch of the systems
but the air from any one room may be diluted 100 to
1000 times, thereby reducing the detectable limit to
10^{-4} to 10^{-5} μCi/cc, above the MPC for Xenon-133,
the major component of the radiogas in purification
and gas handling processes.

The utilization of a portable tritium monitor for
surveillance of work locations subjected to these unsuspected leakages is described as well as its employment for the location of the leaking components.

This instrument lends itself to the measurement
of any beta emitting radiogas and its sensitivity

increases as a direct ratio of the beta energy with a measurement sensitivity for Xenon-133 of 5×10^{-8} microcuries per cubic centimeter.

INTRODUCTION

Most monitoring for radioactive gases is accomplished at nuclear plants by the use of fixed monitors sampling the discharge lines of ventilating systems. Sensitivity is in the order of 1×10^{-7} $\mu Ci/cc$ and is sufficient for monitoring of releases to the environment.

Because of the number of cells or rooms being ventilated and the variations in exhaust air flow from these locations, we have the possibility of gas concentrations in any one cell 100 to 1000 times higher than the minimum sensitivity of the ventilating system monitor. This then could exceed the MPC_a values. Fig. 1 shows the basic ventilating system at Indian Point.

At a pressurized water reactor installation such as the one we have at the Indian Point Station, the major radiogas component is Xe-133 with half life of 5.3 days and a MPC_a of 1×10^{-5} $\mu Ci/cc$.

In most cases only radiogas is present without accompanying radioactive particulate matter. This rules out the more popular and well known air particulate monitors that most reactor installations have.

The Xe-133 is present in the purification and gas handling systems in concentrations varying from 1×10^{-4} to 1×10^{0} $\mu Ci/cc$. This of course allows small leaks in any system to place the concentration in any cell at MPC_a or higher.

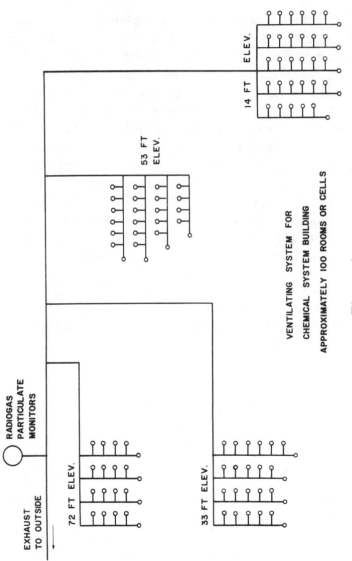

Figure 1

INSTRUMENTATION

The instrument used at the Indian Point Nuclear
Station is the Model 855 TRITON manufactured by the
Johnston Laboratories Inc. of Baltimore, Md. It is
an ionization chamber instrument having four chambers of 5.6 liters apiece utilizing the outside 2
chambers gamma field compensation and the inner 2
chambers for the measurement of the radioactive gas.
In order to keep unwanted particulate matter from
the measuring chambers, a 0.5 micron membrane filter is used as the first stage in the sampling line,
followed by an electrostatic precipitator (Fig. 2).
The measured gas then enters the measuring chambers
free from particulate except for any daughters manufactured in the chambers. The sensitivity is
1×10^{-5} μCi/cc for tritium at full scale indication
on the lowest most sensitive range. The instrument
sensitivity to other radiogases with higher energies than the tritium beta energy is discussed further on this paper.

MEASUREMENTS FOR TRITIUM

With the refueling of the reactor eminent, we
felt the need of an instrument for the detection and
measurement of tritium in air to back up calculations
of possible tritium concentrations due to the primary
coolant being exposed when the reactor head was
removed. We obtained a Model 855 TRITON in time
to be used when Core A was removed and Core B was
installed. This occurred starting in December 1965.
The concentration of tritium in the primary coolant

Figure 2.
The TRITON protable gas monitor in use.

was approximately 1×10^{-1} μCi/cc. The maximum
concentration of tritium in air obtainable from the
primary coolant at 100% Relative Humidity and 80° F
was 8×10^{-6} μCi/cc which is in excess of MPC_a.
With the primary coolant exposed, the TRITON showed
that the concentration in air of tritium was below 10%
of MPC_a and no noble gases present.

DETECTION OF XE-133 AND THE
TRITON SENSITIVITY TO RADIOGASES

We have had tritium concentrations in contain-
ment of 5×10^{-6} μCi/cc, but along with this was the
overwhelming presence of Xe-133 in concentrations
of 5×10^{-4} μCi/cc or 50 times MPCa. The determinations for these concentrations had been made by other
standard test methods and the one for tritium identification was too time consuming. The tests did help
prove the ever present Xe-133 along with other isotopes and future testing showed the presence of Xe-133
only in some systems. The use of the Model 855
TRITON (Fig. 3) now gave us a tool for positive indications of gas activities less than MPC when the monitor
indication is at or below 5×10^{-6} μCi/cc which is half
scale on the most sensitive range.

The instrument can be used to detect all radiogases with a beta energy greater than tritium but the
concentration as indicated on the instrument is not
correct for these other gases. Because the instrument works on the ionization chamber principle with
the sampled air going into the measuring chambers,
the higher the beta energy the greater the number of
ion pairs created by each nanocurrie and therefore

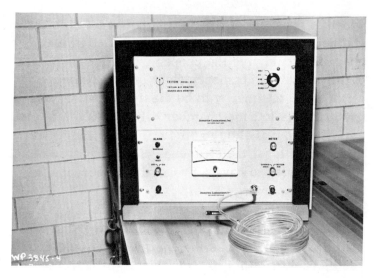

Figure 3
The TRITON portable gas monitor

the higher the indicated concentration per nanocurrie.
While the instrument indication should be directly
proportioned to the ratio of the beta energies of the
gas to tritium, this is in fact not wholly true because
of such factors as the volume of the ionization chamber not allowing complete dissipation of the higher
energy betas.

The volume of the two measuring ionization
chambers that are used in the Model 855 TRITON are
5.6 liters each and the physical dimensions of each
are 8.5 x 20 x 33 cm. The indicated concentration of
the gas being monitored is proportional to the beta
ratios giving us a minimum sensitivity for Xe-133 of
5×10^{-8} μCi/cc. In practice we have found that our
minimum sensitivity on the TRITON for Xe-133 is
approximately 2.0×10^{-7} μCi/cc, which is 2.0% of
MPC. This factor of 5 of course is smaller than 20

which is the ratio of the Xe-133 beta energy to the
tritium beta energy. The straight ratios of beta
energies would give the proper indication increases
if the ionization chambers were of an infinite size to
allow all of the energy to be dissipated before the
beta particle left the chamber. Even a 0.1 MeV beta
particle can travel up to 15 cm in air. Therefore, at
this energy level the efficiency of the chamber is not
what is needed to have a proportional increase in
sensitivity to energy level.

Testing that has been done by the Johnston Company and Consolidated Edison shows a trend at level-
ing off of sensitivity as the energy of the measured
gas increases. Figure 4 graphically illustrates
the theoretical versus practical curves. The effective ratio seems to be between 4 and 6 to 1. Consolidated Edison's check with Xe-133 gave a 5 to 1
ratio and Johnston Company's check with Krypton-85
approximately 5 to 1. The Xe-133 energy is 0.35 Mev
and Krypton-85 is 0.67 MeV. We had intended to
check out the sensitivity using the KR-85 before this
paper was presented but were unable to obtain this
radiogas in time.

Tests have been made by a third party[1] using
Argon-41 gas in a Model 755B TRITON having 3.2
liter chambers and gave a sensitivity ratio for Argon-
41 to tritium of 5.9. The slightly larger chambers in
the Model 855 of Consolidated Edison's TRITON should
increase this ratio only slightly. From the sensitivity
results using Xe-133 (0.35 MeV), Kr-85 (0.67 MeV)
and Ar-41 (1.2 MeV) an educated conclusion can be
reached that from at least the 0.3 KeV level for beta
emitting radiogases the sensitivity factor will not
increase much above the 5 to 1 ratio. Alpha emitting

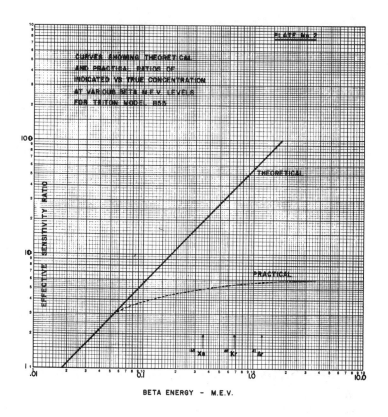

Figure 4

TRITON 855 sensitivity as a function
of beta energy.

gases present a different picture and will be discussed
further on.

We have had one major radiogas problem due to
the escape of Xe-133 where our TRITON 855 proved
its value. This occurred on our Low Pressure Sweep
Gas System which is basically an air blanket over a
number of waste collection tanks. The blanket pressure dropped and the leak did not indicate itself on the
fixed monitoring of the ventilating system servicing

that area. The TRITON located the leak and measured
the concentration in the cell where the leaking pipe
was located. Backup verification was performed by
gas samples checked quantitatively and qualitatively
using a 3" x 3" crystal and a 400 channel analyzer.
After the leak was repaired, the TRITON was used
to ascertain that there was no leak from the repaired
pipe.

DETECTION AND MEASUREMENT
OF RADON-222

While looking for possible leaks, we came upon
another natural phenomena, the presence of Radon-222
radiogas. Having known about the presence of this
radiogas in the environs and in the plant proper under
certain conditions such as a temperature inversion,
we had not seen or even tried to monitor the Radon-
222 with the TRITON 855. We first had an indication
from Radon-222 in the cell where the water from the
containment foundation sump passes before leaving
the plant to the Hudson River. The foundation water
contains Radon-222 coming from the ground deposits
of uranium. Indication on the TRITON was 3×10^{-6}
$\mu Ci/cc$. Other testing confirmed that the air activity
was due to Radon-222 and the concentration due to the
daughter was 2.5×10^{-9} $\mu Ci/cc$. No testing was done
to verify the parent-daughter equilibrium but due to
the small movement of air in the cell it was felt that
the parent and daughters were at or close to equilibrium. Using the concentration for the Radon-222 at
2.5×10^{-9} a sensitivity ratio was obtained for Radon-
222 to tritium of 1200/1. These results compare

favorably to the work done by Waters[2] of the Johnston Laboratories, Inc.

CALIBRATION

Calibration of the TRITON can be accomplished in a number of ways. The Johnston Company manufactures a tritium calibrator enabling the TRITON to be calibrated with tritium gas. Most reactor installations have the means by gamma spectroscopy to identify gas in a system quantitatively and qualitatively. This gas can be used to calibrate the TRITON. Krypton-85 with a half life of 10.3 years is easily obtainable and is useful in checking the calibration for most of the radiogases (Fig. 5).

Figure 5
The TRITON being calibrated with ^{85}Kr

IMPLEMENTAL PROCEDURES

After the initial phase of investigative work using the TRITON Gas Monitor it became standard procedure at Indian Point to use the instrument to monitor all locations where personnel were working in cells that might have leaks from the purification or gas systems. We have been able to feel sure that there are no exposure in excess of 10% of MPC and in most cases to limit this to less than 1% of MPC. With the increased use of the TRITON Monitor, a second unit was purchased for backup. We have encountered numerous situations where the work location had the presence of Xe-133 as shown by the TRITON without any indication on the ventilating system monitors.

CONCLUSION

Portable gas monitors are at least as necessary as particulate monitors for use in the normal nuclear plant operations. Along with these monitors is the need for isotopic identification by spectral analysis in order to interpret what the gas monitor is indicating. The gathering of information for this paper has clearly brought to our minds the importance of this work and the need in the future of more extensive work in this field.

REFERENCES

1. Data supplied by individual, while considered
    good was not substantiated by a calibration of the
    instrument using tritium.

2. J. R. Waters, Calibration of Instrument For
    Measurement of Radon Concentrations in Mine
    Atmospheres, Johnston Laboratories, Inc.
    paper JLI-501B.

EFFECTIVENESS OF FILTER MEDIA FOR SURFACE COLLECTION OF AIRBORNE RADIOACTIVE PARTICULATES*

Dale H. Denham
Pacific Northwest Laboratory
Battelle Memorial Institute
Richland, Washington

ABSTRACT

The relative surface collection efficiencies of cellulose, cellulose-asbestos, and glass fiber filters were determined for alpha activity by comparison with a membrane filter reference.

Alpha activity determinations made from air samples collected on cellulose filters may be less than

---

*This paper is based on work performed under the auspices of the United States Atomic Energy Commission.

1093

one-half of the activity determined from these same
samples by dissolution and alpha counting of the resultant solution. Beta activity determinations made
on the downstream side of air sampling filters yielded
results as low as 50% of those determined from the
upstream side of the filter. Glass fiber filters were
found to have 15 percent greater surface collection
efficiency than the popular cellulose-asbestos filters.

INTRODUCTION

Health physics air sampling programs generally
require sampling for more than one radionuclide in
a given work space. Work with the highly radiotoxic
alpha-emitting materials (such as plutonium) is often
carried out in a separate facility, but other materials
(e.g., fission products) may be present as contaminants or by-products of the major radionuclide being
processed. The filter medium used for sampling in
such a laboratory must retain a consistent and major
fraction of the activity passed through the filter. At
the same time, any alpha emitters collected must be
retained on the filter surface where they can be detected by routine counting.

The following assumptions are frequently made
regarding the effectiveness of air filters to determine
the airborne activity at the location sampled: 1) the
filter collects a constant percentage (usually taken to
be between 90 and 100%) of the particulate matter in
the airstream sampled, 2) the distribution of the
radioactive material collected on the filters does not
vary from sample to sample, and 3) that comparative
measurements are not affected by factors such as

particle size or the shielding of the filter material itself.

Under routine air sampling conditions, however, these assumptions are seldom, if ever, true. Normally, a measured volume of air is passed through a given filter and then the filter is counted for alpha and/or beta activity. The counting rate is converted to a disintegration rate taking into account the geometry and efficiency of the detector-source arrangement and the specific nuclides involved. Then the airborne concentration of radioactive particulates is determined from the volume of air sampled. Because the data collected from such routine air sampling programs is often accepted at face value, the study reported here was undertaken to evaluate the errors associated with this type of analysis.

One of each of four filter types most commonly employed for analysis of airborne radioactivity— cellulose, cellulose-asbestos, glass fiber, and membrane—were used in this study. Air sampling filter media consist of fibers which collect particulate matter by impaction, interception, diffusion, or electrostatic attraction.[1] The membrane filters are synthetic materials with a controlled pore size. Membrane filters trap particles larger than the pore diameter by screening and trap smaller particles as a result of electrostatic and impaction processes. Cellulose (i.e. standard chemical) filters are widely used in air sampling because of their low ash content and ease of chemical destruction, low resistance to air flow, and general ruggedness. Membrane filters are also easy to dissolve and have extremely high collection efficiency, but are fragile, expensive, and have a high resistance to flow. The cellulose-asbestos and glass fiber filters suffer from the fact they cannot be readily dissolved and have high ash content, but otherwise are quite suitable for the routine collection of airborne radioactive particulates.

Some characteristics of these filter types are listed in Table I.

EXPERIMENTAL PROCEDURE AND RESULTS

Samples for the alpha surface collection efficiency experiment were collected within the lab; those for the other tests were collected at outdoor locations. All filters were analyzed using a low background, gas-flow proportional counter with counting backgrounds of approximately 0.2 cpm for alpha and 5 cpm for beta.

Alpha Surface Collection Efficiency

Paired samples of atmospheric dust were collected simultaneously on 47 mm filters. One sampler contained a membrane filter used as the reference collector and the other sampler, operated in parallel, used a test filter collecting medium. Millipore SM (5.0 μ pore size) was used "smooth" side up as suggested by Lindeken[2] and offered a suitable, but not absolute, reference standard. Air was drawn through the filters at 2 cfm ($\pm$ 5%), equivalent to a face velocity of 68 cm/sec. This testing method compares with that of Lindeken[3], except that samples in this study were collected over 12- to 72-hr periods rather than the brief periods used by Lindeken and others.[1-5] The longer sampling times are more typical of routine air sampling conditions, and collection equilibrium

Table I

Characteristics of Four Types of Air Sampling Media

Filter Type	Pressure Drop (mmHg) at 35 cm/sec[1]	DOP (0.3μ) % Pentration at 35 cm/sec[1]	Relative Cost
Cellulose	25	20	1
Cellulose-asbestos	45	1	4
Glass fiber	20	< 0.1	3
Membrane			
Large pore (3-10 μ pore size)	30	-	18
Small pore (< 1 μ pore size)	100	< 0.1	18

for thoron daughters (the time when the number of atoms decaying is equal to the number of atoms being collected) is achieved with the longer sampling times. No backup filters were used here since only surface collection efficiency for alpha-emitters was being studied.

After collection, filter sample pairs (the reference and the test filter) were alpha counted periodically in the low-background system to follow the decay of the radon-thoron daughters for a 2-day period. Counting times were 1-5 minutes during the first few hours after filter removal, then 20-40 minutes for the remainder of the decay period. The counting data were graphed on semilog paper from which the relative surface collection efficiency was determined by a comparison of the alpha counting rate on the test filter with that on the membrane reference filter at the same decay time.

The observed alpha surface collection efficiencies are presented in Table II as the average of several measurements obtained under different atmospheric concentrations and different sample collection periods. The surface collection efficiency of filters is known to be dependent upon flow rate and from this study, the efficiency also appears to be dependent upon sample volume.

The increased efficiency of cellulose filters as compared to the data of Lindeken is probably due to filter loading and hence increased collection. An "apparent" decrease in surface collection efficiency with sampling time (volume) would be expected for long-lived alpha emitters if dust loading on the filter occurred between the time of collection of the alpha activity and the end of the sample collection period.

Table II

Alpha Surface Collection Efficiency

| | Surface Collection % | |
Filter	This study (68 cm/sec)	Lindeken[3] (100 cm/sec)
Cellulose	75	38
Cellulose asbestos	78	80
Glass fiber	92	89

The glass fiber filters used in this study had a better alpha surface collection efficiency than the popular cellulose-asbestos filters. In addition, Phillips and Lindeken[6] showed that RaA (5.99 Mev) and RaC (7.68 Mev) were resolved in pulse-height spectra of natural activity collected on glass fiber (Gelman E) but not on cellulose-asbestos (HV-70) filters confirming the better efficiency with glass fiber filters.

Alpha Activity Absorption in Filter Mat

The reduction in alpha counting rate, because of the penetration of the sampled radionuclides into the filter mat or because of dust covering the previously collected activity, was studied using cellulose-type filters. All samples were collected in sheltered outdoor locations over two- or three-day periods. The samples were collected on 20 cm x 25 cm filters at an average rate of about 30 cfm, equivalent to a face velocity of 35 cm/sec.

Cellulose-type filters were chosen for this environmental sampling program because low ash

content, low flow resistance, and ruggedness were
required. After sample collection, the filters were
cut in half, lengthwise, to permit routine analyses
unassociated with the study. A 10 cm diameter section was cut at random from each of these "half"
filters, placed in a separate glassine envelope, and
retained for one week to permit the decay of natural
daughter product activity.

The week-old 10 cm discs were counted for alpha
activity. Counting time was determined by the time
required to register a minimum of 500 counts. These
10 cm filter sections then were dissolved in acid,
heated to dryness, and redissolved. This resultant
solution, which was assumed to contain all of the
activity from the filter mat, was plated out on 10 cm
steel planchets. The solution was placed on the planchet in a large number of carefully spaced droplets
to provide uniformity and to minimize self-absorption
of alpha particles caused by salt accumulations.
Finally, these planchets were dried and counted, heated
to a bright red with a Bunsen burner, and recounted.
A uranium standard was counted with each batch of
samples to take into account any day-to-day variations in counting efficiency.

Two ratios were computed from the data; the
activity on the 10 cm filter specimen was compared
to that on the planchet both before and after flaming.
The mean ratio in both cases was approximately the
same, indicating about 40% loss of alpha counting rate
due to penetration of collected activity within the
cellulose filter medium or "burial" of the collected
activity by dust on the filter surface, in agreement
with the results obtained by Lindeken. [3] Those samples with apparent high dust loading (i.e., lowest ratio

of alpha activity on the filter to that on the flamed planchet) showed the greatest difference in the two ratios. A possible explanation is that the solids on the planchet, which could mask the plated alpha activity, were removed during the flaming, thereby yielding a higher final alpha count and a lower filter-to-planchet alpha counting ratio. Blank filters and planchets also were evaluated to determine the net alpha activity from the filter.

For samples collected during a given period of time (e.g., one week), the standard deviation of the mean (in units of % alpha loss) was only $\pm$ 10%. However, the deviation from the mean of all the data collected during several months was nearly $\pm$ 40%. Examination of the results of the air sampling, which was conducted during variable atmospheric conditions ranging from "dust" storms to relatively clear, led to the conclusion that dust loading was a major factor in determining the apparent loss of alpha activity within cellulose-type filters. In general, the observed alpha activity on cellulose-type filters was only about 50% of that found by dissolution of the filter and counting of the resultant solution.

Beta Absorption in Filters

Finally, the self-absorption of beta activity deposited on and within the filter media used in this study was also investigated. Atmospheric samples containing principally world-wide fallout (long-lived fission products) were collected on 20 cm x 25 cm filters using a multistage centrifugal exhauster. Average flow rates were between 20 and 30 cfm, equivalent to face velocities through the filter in the order of 28 cm/sec. Samples were collected over either 24 or 48 hour periods, then permitted to decay for 7 days prior to counting to minimize the contribution of naturally occurring nuclides to the total activity.

Square sections of approximately 60 cm^2 each were randomly cut from these filters. Each square (not more than two were taken from any given filter) was weighed to the nearest milligram on a Mettler single pan balance. These filter squares were counted in the low-background system, both face up and face down to give an indication of the reduction in beta count rate through the filter. (Alpha and beta activities are determined on our routine air sample filters by counting simultaneously for alpha on the upstream side and beta on the downstream side of the filter.)

The absorption of beta activity in the filters was found to be dependent upon the total mass of the "loaded" filter in agreement with the findings of Lockhart,[4] et al. The depth of penetration of collected nuclides within the filters was not determined. The average ratio of beta counting rate from the downstream and upstream sides of each filter type is presented in Table III.

No consistent mathematical relationship between the observed beta activity ratios and the mg/cm^2 of the "loaded" filter was found. The data point out, however, the varying nature of atmospheric radioactivity and the difficulty in assessing airborne concentrations based upon a gross beta count.

Table III

Average Beta Absorption in Filter Media

Filter-type	Density-thickness of "Loaded" Filter (mg/cm^2)	Net Beta Counting Ratio (back/front)	
		Average	Range
Cellulose	9.7	0.84	0.71-0.92
Cellulose-asbestos			
9 mil	8.7	0.78	0.76-0.80
18 mil	17.6	0.63	0.61-0.67
Glass fiber	5.7	0.72	0.52-0.93
Membrane (small pore)	5.3	0.85	0.75-0.99

SUMMARY

A series of "field-type" experiments were performed to evaluate the results of routine air sampling programs for radioactive particulates. Several conclusions may be drawn from the results of the tests.

Cellulose-type filters should not be used for quick, high-volume, grab sampling in a plutonium atmosphere because of the significant loss of alpha activity by penetration into the filter media or by being covered up by dust collected on the face of the filter.

Because of greater surface collection efficiency, lower resistance to air flow, and lower price glass fiber filters are superior to the popular cellulose-asbestos types for routine gross radio-activity analysis.

Arbitrary use of single correction factors to account for filter collection efficiency, penetration of sampled materials into the filter, or self absorption of filter media may result in the underestimation of the true concentration of radioactivity in the air sampled.

ACKNOWLEDGMENTS

The author wishes to express appreciation to C. L. Lindeken and the Radiation Safety Counting group at the Lawrence Radiation Laboratory in Livermore, California for their assistance in setting up these studies and in analyzing and storing the samples. Thanks are also due to D. W. Alton for his helpful discussions and comparison of similar data.

REFERENCES

1. L. B. Lockhart, Jr., R. L. Patterson, Jr., and W. L. Anderson, "Characteristics of air filter media used for monitoring air-borne radioactivity", NRL-6054 (1964).

2. C. L. Lindeken, "Surface collection efficiency of large-pore membrane filters", UCRL-7254, pp 16-27 (1962).

3. C. L. Lindeken, Amer. Industr. Hygiene Assoc. J. 22, 232 (1961).

4. J. J. Fitzgerald, and C. G. Detwiler, "Collection efficiency of air cleaning and air sampling filter media", KAPL-1088 (1954).

5. S. Posner, "Air sampling filter paper retention studies using solid particles", in USAEC Report No. TID-7627, (1961).

6. W. A. Phillips and C. L. Lindeken, Health Physics 9, 301 (1963).

THE TWO-FILTER METHOD
FOR RADON-222

Jess W. Thomas and Philip C. LeClare*
Health and Safety Laboratory
U. S. Atomic Energy Commission
New York, N. Y.

ABSTRACT

The two-filter method for radon-222 has been improved so that it is rapid and simple, requiring only ordinary alpha counting. Radon containing air is passed through two filters in series separated by a known cylindrical volume V. The first filter passes radon but removes all daughter activity. The second filter collects daughter products formed in flight in the volume V. The activity of radon (C_{Rn}) in picocuries per liter is given by $(0.450X)/EZVF_f$ where X is the number of alpha counts on the exit filter for a given sampling time and counting time, E is the counter efficiency, Z a decay correction factor,

---

*Presently a doctoral candidate at Rutgers University, New Jersey.

and F_f the fraction of radium-A atoms formed in V that arrives at the exit filter.

Tables of Z for various sampling and counting times have been developed. Likewise, values of F_f are given from solution of the equation for diffusion in a cylindrical tube with formation in flight. Derivation of these Z and F_f factors makes the method absolute and theoretically no calibration is required.

The sensitivity of the method is of the order of a few picocuries per liter for sampling and counting times of about 15 minutes and a tube volume of 1/2 liter. Much higher sensitivities are possible for large tube volumes and long sampling and counting times.

INTRODUCTION

The two-filter method for radon-222 has been studied since the early 1960's with important papers appearing by Fontan[1] and Jacobi.[2] Briefly, air to be assayed is drawn through two filters in series, which are separated by a delay volume to permit radon decay. The radon concentration is calculated from the alpha activity of the radon daughters caught in the downstream filter. Figure 1 shows a typical arrangement in which a cylindrical tube 50 cm long and 3.6 cm inside diameter provides the delay volume. It is necessary to make corrections for decay and diffusion loss, and the principal deficiency of all work to-date has been the lack of accurate expressions for these corrections. The present work corrects this deficiency.

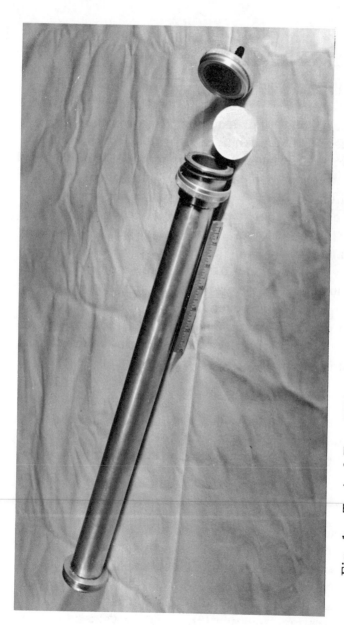

Fig. 1. Typical Two-Filter Tube With Detached Exit Filter Holder and Filter.

THEORY

Basic Radon Concentration Equation

The equation is derived for short tube transit times, that is, it is assumed that the decay of radium A is negligible. For example, a typical tube with a volume of 500 cc, might be operated at a flow rate of 10 liters per minute. The transit time would be about three seconds, compared to the 3.05 minute half-life of radium A. *

Let Q_1 be the concentration of radon, atoms per liter, in the tube and in the entering air, λ_1 the decay constant of radon, dis/(atom) (min), V be the tube volume, liters, and q the flow rate, liters per minute. The number of radon atoms in the tube at any time is then $Q_1 V$, which produce $Q_1 V \lambda_1$ atoms of radium A per minute. Since the fraction of the radium A atoms reaching the filter is F_f, the number of atoms of radium A arriving at the filter per minute is

$$Q_1 V \lambda_1 F_f = \Phi \tag{1}$$

If we let C_{Rn} be the radon concentration in picocuries per liter, then $Q_1 = 2.22\, C_{Rn}/\lambda_1$ and $\Phi = 2.22\, C_{Rn} V F_f$ Solving for the radon concentration C_{Rn} gives

$$C_{Rn} = \frac{0.450\,\Phi}{V F_f} \tag{2}$$

In equation (2), the tube volume V is known. Quantity Φ, which is the number of radium A atoms arriving at the filter per minute, may be obtained from alpha counting of the exit filter. F_f may be obtained from diffusion theory. Thus, the basic equation radon concentration, equation (2), requires values for Φ and

---

*Radium A is ^{218}Po

F_f, which in turn are calculated from the tube dimensions, flow rate, and sampling and counting times.

Values for Quantity Φ

To obtain values for Φ, it is convenient to define a factor Z as follows;

$$Z = X/E\,\Phi$$

where X is the number of alpha counts obtained for a sampling time t and counting interval T_1 to T_2, where T = 0 at the end of sampling. Quantity E is the counter efficiency. From an analysis of the buildup and decay of activity on the exit filter, a set of values of Z has been obtained by means of a computer, as a function of t, T_1, and T_2 (Table 1). From these values of Z, quantity Φ may be obtained, since $\Phi = X/EZ$.

Values for Quantity F_f

The standard Gormley-Kennedy equations[3] give the fraction F_p of entering atoms not diffusing out to the wall of a cylindrical tube as a function of flow rate q, diffusion constant of the diffusing species D, and length of tube, L;

$$F_p = 0.819 \exp{(-3.66\mu)} + \ldots\ldots\ldots$$

where $\mu = \pi\,DL/q$.

For the case of radon decay during flow through a tube, however, the situation is different. Instead of the diffusing atoms all entering the inlet of the tube, they form "in flight" throughout the entire tube volume. This necessitates rewriting the basic differential equation and resolving the equation, which has been done recently by C. W. Tan[4]; values of F_f, the

Table 1.

Values of Z as a Function of t, T_1 and T_2

t	T_1	T_2	Z
5	1	6	1.672
5	1	15	2.597
5	1	30	3.411
5	1	100	6.314
10	1	6	2.312
10	1	15	3.803
10	1	30	5.425
10	1	100	11.068
15	1	6	2.656
15	1	15	4.634
15	1	30	7.070
15	1	100	15.281

Z = decay correction factor, min.

t = sampling time, min.

T_1 = start of counting period, min.

T_2 = end of counting period, min.

(T=0 at end of sampling)

fraction of atoms formed in the tube that reaches the exit filter, are given as a function of μ.

In using Table 2, the diffusion constant of radium A is to be taken as 0.085 cm^2/sec. This is the approximate value found in preliminary investigations at this laboratory.

APPLICATION OF THE METHOD

Working Equation and Sensitivity

Equation (2) is most conveniently applied by substitution of $\Phi = X/EZ$, which gives the working equation

$$C_{Rn} = \frac{0.450X}{EZVF_f} \tag{3}$$

Quantity X is determined by counting the exit filter, E is found by use of an alpha standard, Z is read from Table 1 as a function of t, T_1, and T_2, V is known, and F_f determined from Table 2 as a function of $\mu = \pi DL/q$, where L and q are known and D is taken as 0.085 cm^2/sec.

The sensitivity of the method is a function of tube diameter, tube length, flow rate, sampling time, alpha background, counting efficiency, and counting interval T_1 to T_2. For example, a tube 3.6 cm diameter, 50 cm long, operated at 10 liters per minute, for a sampling time t of 15 minutes, with T_1 = 1 min. and T_2 = 16 min. (15 min. counting time) may be used to detect 10 picocuries per liter with a standard deviation of about 10-20%.

Table 2.

Values of F_f as a Function of $\mu = \pi DL/q$

μ	F_f	μ	F_f
0.005	0.877	0.25	0.420
0.008	0.849	0.30	0.391
0.01	0.834	0.35	0.349
0.02	0.778	0.40	0.324
0.03	0.737	0.45	0.302
0.04	0.705	0.50	0.282
0.05	0.678	0.60	0.248
0.06	0.654	0.70	0.220
0.07	0.633	0.80	0.197
0.08	0.614	0.90	0.178
0.09	0.596	1.00	0.162
0.10	0.580	1.50	0.110
0.12	0.551	2.00	0.083
0.14	0.525	2.50	0.067
0.16	0.502	3.00	0.056
0.18	0.481	4.00	0.042
0.20	0.462	5.00	0.033

F_f = fraction of atoms formed in tube arriving at exit filter

π = 3.1416

D = diffusion constant, 0.85 cm^2/sec for radium-A

L = tube length, cm.

q = flow rate, cm^3/sec.

Practical Precautions

Leaks, even if small, cannot be tolerated in this method, as the method depends on counting the radium A formed in flight in the tube, which is usually far less than the radium A normally present in a mine atmosphere. Hence a small leak, for example, around the entrance filter will give falsely high radon concentrations. Operation of the two-filter tube under slight positive pressure is recommended so that any small leaks are outward, rather than inward.

Likewise, precautions must be taken with the filters. Filters likely to be charged, such as "Millipore" filters, are unacceptable as they accumulate activity from the mine atmosphere. Type 934 AH glass fiber filters are recommended, and these should be stored in individual glassine bags.

REFERENCES

1. J. Fontan, "The Quantitative Determination of Gaseous Radioelements which Yield Radioactive Daughter Products," PhD Thesis, University of Toulouse, Transl. by R. R. Inston, Argonne National Laboratory, ANL-TRANS-45 (1964).

2. W. Jacobi, "A New Method to Measure Radon and Thoron in Streaming Gases and its Use to Determine the Tn-Content of Atmospheric Air," Proceedings of Int'l Conference on Radioactive Pollution of Gaseous Media, Saclay, France, Nov. 11-16, 1963, Presses Universitaires de France, p. 509 (1965).

3. P. G. Gormley and M. Kennedy, Proc. Roy. Irish Acad. $\underline{52}$, 163 (1949).

4. C. W. Tan, "Diffusion of Disintegration Products of Inert Gases in Cylindrical Tubes," Accepted for publication in the Int. J. of Heat and Mass Transfer (1968).

PLUTONIUM SENSITIVE ALPHA AIR MONITOR*

C. E. Nichols
Health and Safety Branch
Idaho Nuclear Corporation
Idaho Falls, Idaho

ABSTRACT

A plutonium sensitive Alpha-Air Monitor, utilizing two ZnS scintillation detectors, has been developed at the Idaho Nuclear Corporation Health and Safety Laboratory that will detect plutonium-239 in the presence of the most adverse natural activity atmospheres found within the National Reactor Testing Station. The instrument, without the use of discriminators, has demonstrated that it can eliminate greater than 90% of the 2,250 d/m alpha natural background activity collected in order to detect 48 RCGs of soluble Pu-239 within 1-1/2 hours. With the use of discriminators, the instrument should be able to detect even lower concentrations of plutonium-239.

---

*Work performed under the auspices of the Atomic Energy Commission.

INTRODUCTION

The detection of airborne plutonium-239 in relatively small quantities becomes a very difficult problem at the National Reactor Testing Station because of abnormally high concentrations of natural background activity. Experience has shown that it is not uncommon to find natural background activity concentrations exceeding 1.8×10^{-8} $\mu Ci/cc$. Idaho Nuclear Corporation, at present, depends on an alpha air-monitor that uses the alpha—beta ratio technique. These particular air-monitors are relatively insensitive to plutonium. It would take a 190 RCG concentration of soluble Pu-239 to give a significant count during a one hour period under inversion conditions when a steady state background reading is achieved.

The Alpha-Air Monitor described in this paper uses two detectors, one a reference detector which compensates for natural background activity, the other being a sample detector, amplifiers, a reversible counter, and associated equipment. The ability of the instrument to detect low concentrations of plutonium, in the presence of relatively high concentrations of natural background activity, is dependent on the digital subtraction of pulses of reference detector from sample detector. The counting statistics are improved by counting for ten minutes during each cycle.

DESCRIPTION OF INSTRUMENT

The instrument is a semi-portable unit with the electronics, the two ZnS scintillation detectors, and the air mover mounted on a wheeled cart. One of the detectors (reference channel) monitors air outside the plutonium handling facility. The second detector (sample channel) monitors the air inside the plutonium handling facility. The actual air sampling is accomplished by drawing air directly from the area being monitored into the sample channel detector. A hose is connected from the reference channel detector to the outside of the plutonium handling facility for monitoring of a plutonium free atmosphere.

The air mover is an integral component of the instrument capable of approximately 5 cfm with each channel operating at 2 cfm. For the initial tests on the instrument, 50 mm H.V.-70 filter paper was used. The detector and air collection arrangement are very similar to the one used at Rocky Flats.[1] The detectors use ZnS phosphor coated directly on the photomultiplier tube. A thin coating of high vacuum grease bonds the ZnS to the surface of the tubes. This method is very satisfactory in that the ZnS is evenly distributed, and during the operation of the instrument the phosphor is not pulled off by the air flow.

Figure 1 is a block diagram of the system. The ZnS scintillation detectors detect the alpha particles, convert to current pulses, which are then amplified by the preamplifiers and amplifiers. After passing through the discriminator circuits, the pulses are applied to the reversible counter inputs.

The reversible counter is of a type which counts in an upward direction (1.2.3....) for pulses received at one input, and downward (....3.2.1) for pulses received at the other input. This function effectively subtracts one input from the other. With this function, the counts from the reference channel can be subtracted from the counts of the sample channel.

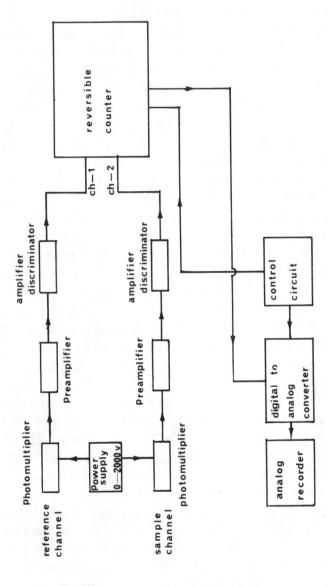

Fig. 1. Block Diagram of Plutonium Sensitive Alpha Air Monitor.

The control circuit allows the reversible counter to count for a fixed period, say ten minutes, to improve the statistics. At the end of this fixed period, the count remaining in the counter is the difference between the sample channel and the reference channel. This digital number is converted to an analog signal and recorded. If the analog signal exceeds a limit set by the operator, an alarm is energized. At the completion of the recording of the count, the control circuit resets the counter and starts a new counting cycle.

OPERATION

The instrument is operated with each channel drawing 2 cfm with the detectors at 30% efficiency as calibrated with a Pu-239 source. The maximum efficiency on one channel was 33% and 34% on the other channel, and the air mover used is capable of approximately 2.5 cfm per channel. The instrument is operated at lower efficiencies and lower air flows to allow the operator to balance the two channels. In the actual operation of the instrument, it was found that perfect balance conditions are not crucial. In fact, it was found desirable to have the sample channel counting slightly above the reference channel in order to obtain positive results on the recorder. As long as the operator knows the location of the baseline (as determined during the calibration procedure), the alarm can be set. This operation can be seen in Fig. 2.

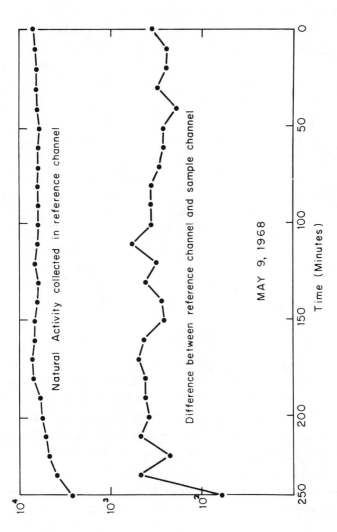

Fig. 2.   Typical Operation of Alpha-Air Monitor with the Natural Activity
          Superimposed on the Chart.

DISCUSSION

For reasons of economy, the instrument described in this paper was fabricated from borrowed equipment and was operated without the use of discriminators. The use of discriminators, to discriminate against the higher energy natural background activities, will most certainly give the instrument a much better sensitivity to Pu-239. The instrument was operated for several weeks with minor air-flow adjustments and periodic filter changes with very good results.

Sensitivity was tested by allowing the instrument to run for several cycles in order to check for proper instrument operation and to establish an operating baseline. Both filters were then removed and the reference channel was supplied with a fresh filter and a "spiked" U-233 filter was inserted in the sample channel. The 200 d/m alpha "spiked" filter was detected almost immediately (first ten minutes). See Fig. 3.

This instrument is relatively expensive compared to less sensitive air-monitors, but the early detection of Pu-239 could prevent a very costly contamination problem to say nothing of the health hazard involved. One such contamination problem at the Test Reactor Area cost several thousands of dollars to decontaminate the facilities.

The instrument that has been described should be used in a building with a common intake air supply because of the possible differences in the natural background activities. If the two detectors were placed in areas with different air supplies, there could be enough of a variation in natural background activity concentrations to impair the sensitivity of the instrument.

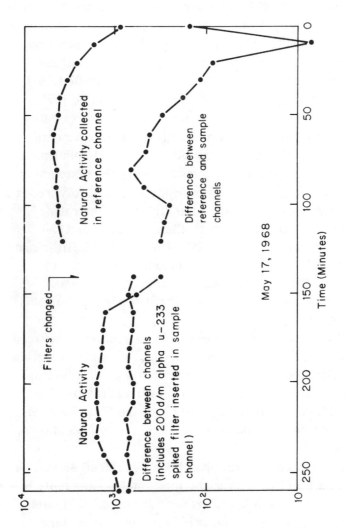

Fig. 3. Detection of a 200 d/m Alpha U-233 Source After Operating the Instrument for a Period of Time.

SUMMARY

The instrument that has been described is capable of detecting 9.6×10^{-11} μCi/cc of soluble Pu-239 in the presence of 1.8×10^{-8} μCi/cc of natural background activity within 1-1/2 hours of sampling. This sensitivity is made possible by the use of two detectors and a reversible counter. One of the detectors collects and monitors the atmospheric alpha activity from the air supply to the monitored area and the second detector collects and monitors the atmospheric alpha activity in the monitored area. The reversible counter subtracts the pulses received from one detector from the pulses received from the second detector. The count difference between the two channels is accumulated for ten minutes of each cycle, to improve counting statistics, and the results are then recorded. Although the described instrument did not use energy discriminators, their use to reject pulses from the higher energy alphas of natural activity would certainly give even better results.

REFERENCE

1. R. A. Kirchner, V. R. Griep, and W. M. Hartzell, "Rocky Flats Continuous Air Monitor", USAEC Report RFP-815 (November, 1966).

AIR MONITORING AND ITS EVOLUTION AT THE LASL PLUTONIUM FACILITY

Allen M. Valentine, Dean D. Meyer
and William F. Romero
Los Alamos Scientific Laboratory
of the University of California
Los Alamos, New Mexico

ABSTRACT

A program to monitor airborne contamination in the working environment of a plutonium facility is discussed. Only those aspects most closely associated with operational monitoring are included and, as a result, several side topics in the realm of air monitoring are not mentioned.

Techniques, equipment, records, and philosophy are discussed, in addition to air concentration trends, equipment evolution, cost, manpower requirements, and inherent shortcomings.

For the present air monitoring program, 351 samples are collected daily on HV-70 filters at fixed locations in work areas and counted in automatic alpha counting systems. Counting results are interpreted in terms of air concentrations and recorded for

evaluation by the health physics and operating staff.
Also, 21 commercial continuous air monitoring devices are located throughout the work areas.

Air monitoring at the Los Alamos Scientific Laboratory (LASL) plutonium facility began in the mid-
1940's when the facility started operation as the
world's first production and fabrication facility for
handling kilogram quantities of plutonium.   The air
monitoring program has since evolved into an adequate
and practical program for accessing airborne contamination levels and for preventing inadvertent inhalation
cases.

INTRODUCTION

A brief review of past techniques and equipment
seems appropriate in that 1968 marked the 25th year
of plutonium handling at Los Alamos.   The first
quantity of plutonium large enough to be seen with the
naked eye arrived in 1943 and plutonium handling increased from a few micrograms in 1943 to kilograms
in 1945.   This left less time than usual for development of safe handling methods and equipment. Production techniques were developed as the need arose
on bench tops and inside conventional chemical fume
hoods.   Meanwhile, air sampling methods borrowed
from other fields were used.   Impingers, electrostatic precipitators, and filter paper samplers were
used; however, filter paper soon gained acceptance for
routine use.   Realizing that it was impossible to take
an air sample which truly represented a worker's
exposure, we began a search for supplemental methods
for evaluating personnel exposure.   This resulted in

the development of a plutonium urinalysis program which began in December 1944, and a nasal swab program in 1945.

The need for better confinement of contamination was soon realized and unventilated wooden dryboxes were introducted. The dryboxes rapidly became standard equipment for plutonium operations that were being conducted in D Building, a large temporary building. In 1946, a new production facility was built which allowed separation of production and laboratory work. This separation proved beneficial because control measures necessary for handling kilogram quantities of plutonium are not always required where microgram quantities are handled. This production facility has undergone several modifications since 1946 and continues to be the main LASL plutonium facility.

Airborne plutonium levels at the facility were reduced significantly between 1946 and 1951 as shown in Fig. 1. This was primarily due to the introduction of ventilated glove boxes and hoods and the elimination of open-air material transfers.[2,3] Equipment and techniques continued to improve and today airborne plutonium is seldom detected in work areas during normal operating conditions.

FACILITY DESCRIPTION

The LASL plutonium facility, known locally as DP West, consists of five buildings with approximately 40,000 square feet of operating area. These buildings are divided into nine wings connected by a spinal corridor which originates at an administration building. Six wings are devoted to plutonium work; the

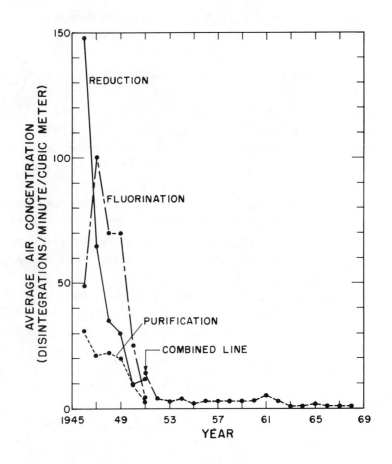

Fig. 1.   Average Air Concentration Trends.

other three are used for enriched uranium recovery
and production and for transplutonium element research.

Sixty-five individuals are directly involved in
plutonium programs which include purification, metal
production, fabrication, recovery, research, and
development. Approximately 200 employees work at
the facility. The plutonium programs involve a wide
variety of operations which are conducted in well-
ventilated and filtered metal glove boxes or enclosures.
Even though kilogram quantities of plutonium are
handled, most of the work is for specialized programs
and is developmental in nature. Only a few programs,
such as purification, metal production, and recovery,
can be classed as production.

MONITORING PHILOSOPHY

The basic philosophy behind air monitoring for
plutonium has not supported the use of air sample results for the primary determination of personnel exposure. However, the philosophy supports an extensive program of sampling with filter paper at fixed
locations throughout the work areas and the use of
gross alpha continuous air samplers for the detection
and warning of excessive levels.

The fixed samples are useful for determining the
effectiveness of operational contamination controls
and personnel exposure in a qualitative manner since
general levels of airborne contamination are measured.

Urinalysis and in vivo counting techniques are
the primary methods of determining personnel exposure for reporting purposes.

PRACTICES AND EQUIPMENT

A. Fixed Sampling

The Filter Queen sampler was used at other Manhattan Project installations prior to becoming the most common fixed sampling system at Los Alamos. The original Filter Queens and later versions were modified household vacuum cleaners (Fig. 2) that pulled air through a 4- x 9-in. filter. They could be moved about the area by one person but were generally left at a fixed location. Their disadvantages included high maintenance cost, the large area of filter paper required for a low pressure drop, noise, and use of floor space. The Filter Queen was used exclusively until 1957 when the installation of central sampling systems (Fig. 3) patterned after the Rocky Flats system was started. Installation of these systems permitted increased sampling in operating areas and the use of small diameter filter papers which could be counted in automatic counting systems.

The central sampling systems utilize 5 high-capacity vacuum exhausters, sized piping to sampling locations in operating areas, quick-disconnect filter holders (Fig. 4), and 2-1/8-in. filters. Air flow through the filter can be adjusted at each location and is checked weekly by the calibrated equipment shown in Fig. 5. A total of 351 locations are used for collecting daily routine samples. The normal sampling period is from 0830 to 1630; filters are replaced during the last 30 min of the work day. The used filters are collected in racks and stored in a clean area until they are counted the following day. Provisions are made for replacing the routine filters during

Fig. 2.   Filter Queen Sampler.

Fig. 3.   Work Area Central Sampling System.

Fig. 4.   Filter Holders and Rack.

Fig. 5.   Flow Calibration

the day and collecting "special" samples whenever operations of a nonroutine nature are conducted. Persons in the area wear respiratory protection during collection of special samples.

Sampling in normally unused areas outside the central sampling system is done with a portable "giraffe" sampler (Fig. 6). This sampler uses a filter holder and filter identical to that used on the central sampling system and is capable of sampling at 2 cfm.

B. Continuous Sampling

Early efforts were made to develop a continuous sampler capable of rapidly detecting excessive airborne plutonium because fixed sample results were not available until a day or so after collection of the sample. Figure 7 shows an early continuous sampler built and used at Los Alamos. It utilized a movable strip filter which advanced from the sample collection position to the scintillation detector every 10 min. When commercial samplers became available they were tried and adopted; 21 continuous samplers are now in use at the facility. Included are samplers from five different manufactors, all with fixed filters and gross alpha scintillation detectors in addition to local audible and visible alarms and recorders.

Most of the samplers are connected to outlets on the central sampling system to reduce pump maintenance and noise. Small samplers are desirable because of the limited space around glove-box lines and the type most widely used is shown in Fig. 5.

Since the samplers utilize fixed filters and gross alpha detectors, they operate with a background count

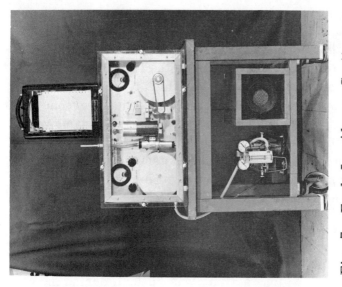

Fig. 7.   Early Los Alamos Continuous Air Sampler.

Fig. 6.   "Giraffe" Sampler.

due to the collection and detection of naturally occurring alpha emitters in the air.   This inherent background makes rapid detection of low levels difficult. Another contributing factor which makes rapid detection of low levels difficult is the fact that the alarm trip must be set sufficiently high to minimize inadvertent or spurious alarms; otherwise, persons in the area may become distrustful of and unresponsive to genuine alarms.   In spite of the difficulties, these samplers have on occasion detected excessive levels which would have otherwise gone undetected until the following day.

C.   Count Room

The task of counting numerous 4- x 9-in. filter papers for alpha activity was difficult during the early years because counting techniques and equipment were in the development stage.   The first satisfactory counters were manual scalers with large gas proportional detectors (Fig. 8).   Each filter had to be loaded manually and each count was recorded by hand.   Operation of the counters required a number of technicians and the results were subject to human error.   Replacement of these counters with ones having automatic sample changer and readout capabilities began in 1964.

Instrumentation in the central count room now includes two Widebeta II systems, two Nuclear Measurement Corporation Model PC-3T counters, and a scintillation counter assembled at Los Alamos.   The Widebeta counters, used for counting routine sample filters, are capable of counting 100 filters per loading and automatically printing identified results on a tele typewriter readout unit (Fig. 9).

Fig. 8. Early Los Alamos Filter Counters.

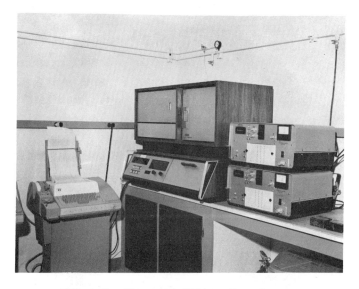

Fig. 9. Present Filter Counters.

Daily precount calibration checks, to determine
the counter efficiency, are made with a standardized
source.   The counting time, approximately 2.7 min,
is used so that the total number of counts accumulated
during the count period will equal the alpha activity in
disintegrations per minute.   This eliminates an arith-
metic step for the full-time count technician.   The
counter efficiency is the only factor applied in converting results to activity in disintegrations per minute
(d/min).

RESULTS AND RECORDS

Once the filter activity in d/min has been determined, the task of preparing permanent records of
the sample must be considered.   A logical way is to
accumulate the results by room in chronological
order.   However, experience has shown some advantages in maintaining an additional personnel exposure
record for each individual.   The following is the se-
quence used to generate this record: (1) each sample
result is identified and converted to a concentration
on Form 1 (Fig. 10) which also includes sample location and time, count time, and sampling rate; (2) the
data on Form 1 are used to determine an average concentration for each room which is recorded on Form
2 (Fig. 11); and (3) the Health Physics Surveyors use
these averages to prepare a Personnel Exposure
Record, Form 3 (Fig. 12) for each person assigned
to their area.
Forms 1 and 2 are completed by the count room
technician and are placed in permanent storage after
one year.   The Personnel Exposure Records are

AIRBORNE CONTAMINATION TESTS
AREA DP West

Room 201 LOCATION	1-9-69 TIME			m³/min	m³	DATE COUNT TIME d/min	d/min-m³	REMARKS
	BEGIN	END	HRS					
1-A	8:00	4:30	8½	.056	28.6	29	1	171.6 M³
1-B						37	1	
1-C						25	1	
2-A						39	1	ave. 1
2-B						31	1	
						̄ ̄	1	74 d/r

Fig. 10.   Individual Sample Record.

AIRBORNE CONTAMINATION TESTS

AREA DP West   LOCATION Room 201

TEST	DATE	TOTAL m³	d/m	d/min-m³	REMARKS
1	1-2-69	429.0	289	1	Group CMB-11 (alpha Pu)
2	1-3-69	"	198	0	
3	1-4-69	"	229	1	
4	1-5-69	"	70	0	
5	1-6-69	"	327	1	
6					

Fig. 11.   Average Sample Record.

PERSONNEL EXPOSURE RECORD
AIRBORNE ALPHA CONTAMINATION

DP WEST AREA

Name: Doe, John Q
Z-00000

January, 1969			(--) Person not in immediate area			
Date	Room No. and Average d/min-m³					Remarks
January	Room 500	Room 513				
2	0	0				
3	0	0				
4	0	0				
5						

Fig. 12.   Personnel Exposure Record.

placed in the individual's Medical Record file at the
end of each year. Because air samples are a measure of operation contamination control, supervisors
in the operating group are notified immediately of
routine sample concentrations in excess of 4
d/min-m^3. These samples are also reported in weekly written reports to the Health Physics Group Leader
and the Operating Division Leader.

Samples collected during normal operations seldom measure concentrations over 4 d/min-m^3; however, over the past five years they have averaged
1 d/min-m^3. This average concentration is 25% of
the 2 x 10^{-12} μCi/cc air concentration standard for
soluble plutonium and is probably caused by natural
occurring alpha emitters remaining on the filter paper
after the overnight decay period rather than plutonium.

COSTS

Costs are briefly mentioned for those health
physicists who are considering such a program.
Estimated initial equipment costs along with manpower and material costs are given in Table I. Manpower and filter paper expenses alone amount to
$0.32 per routine sample. Expense items not in-
cluded are record forms, equipment maintenance,
incidental equipment replacement, and record storage.

SUMMARY

The present program for monitoring airborne
plutonium in the LASL plutonium facility working

Table I.

Estimated Manpower, Material and Initial
Equipment Costs.

Manpower		351 Samples per Day
Health Physics Surveyors (5 h/day @ $7.50/h)		$0.11/sample
Count Technician (8h/day @ $7.50/h)		0.17/sample
Filters		
HV-70, 2-1/8-in. diam		0.04/sample
	Total	$0.32/sample
Equipment		Initial Cost
Counting room counters		$35,000
Continuous air samplers		32,500
Central sampling systems		30,000
	Total	$97,500

environment has evolved from early equipment and
techniques into an adequate and practical system.
Central sampling systems and automatic filter counters
permit sampling at more than 350 fixed locations.
Samples collected during normal operations seldom
measure concentrations over 4 d/min-m^3; however,
over the past five years they have averaged 1
d/min-m^3. This concentration over 25% of the 2 x
10^{-12} μCi/cc air concentration standard for soluble

plutonium. This would be a significant shortcoming in the program if the user wanted to prove that air concentrations or chronic exposure levels were below 1 d/min-m^3.

The general air concentration trend has been downward and experience has shown that the urinary excretion levels do not exceed detection limits for present-day chronic exposures received by workers at the facility. Measurable exposures from airborne plutonium occur only under accidental conditions.

The program is adequate for the routine monitoring of airborne plutonium; however, additional data must be obtained before a true evaluation of health hazards can be attempted.[4,5] Personnel exposure is difficult to determine from the air monitoring results, and techniques directly involving the individual must be relied on for making final exposure evaluations.

ACKNOWLEDGMENTS

The authors are grateful to Julia Emery for typing and arrangement of the final report.

REFERENCES

1. H. F. Schulte and D. D. Meyer, "Control of Health Hazards in Handling of Plutonium: Results of 14 Years' Experience," Proc. U. N. Intern. Conf. Peaceful Uses At. Energy, 2nd, Geneva, 1958, vol. 23, 206-210.

2. W. R. Kennedy, "Problems of Air-Borne Contamination in Handling PuF_4 Powders," Los Alamos Scientific Laboratory Report LAMS-596, (1947).

3. W. R. Kennedy, "Studies of Air-Borne Contamination Resulting from Operations Handling Plutonium Metal and Plutonium Alloys," Los Alamos Scientific Laboratory Report LAMS-725, (1948).

4. W. D. Moss, E. C. Hyatt, and H. F. Schulte, "Particle Size Studies on Plutonium Aerosols," Health Phys., 5, 212-218, (1961).

5. W. H. Langham, J. N. P. Lawrence, Jean McClelland, and L. H. Hempelmann, "The Los Alamos Scientific Laboratory's Experience with Plutonium in Man," Health Phys. , 8, 753-760 (1962).

AN AUTOMATIC INCREMENTAL
AIR SAMPLER

Richard P. Marquiss
Naval Radiological Defense Laboratory
San Francisco, California

ABSTRACT

An automatic incremental air sampler has been developed to sequentially assess the concentration of airborne radioactivity associated with a continuing series of experiments with short-lived early time 235U fission products.

Essential characteristics of the sampler are:
(1) continuous automatic incremental operation,
(2) inexpensive GM detectors, (3) selectable sampling intervals, (4) selectable decay counting intervals,
(5) digital scaling and print-out, and (6) "portability" and remote sampling characteristics.

Examples of the use of this unit and some suggested improvements are presented.

INTRODUCTION

During 1962, the Naval Radiological Defense Laboratory instituted a program of nuclear reactor irradiation of various materials for activation analysis and for study of some of the short-lived fission pro-products of ^{235}U.

While the details of these experiments are not germain to the subject of this presentation, a brief description of the nature of the experiment and the apparatus used will aid in understanding the development of the air sampler.

Most of the experiments consisted of the irradiation of small amounts of fully enriched ^{235}U to about 10^{10} to 10^{13} fissions. The irradiated samples, usually liquid, were contained in an "inert" rabbit within a gas operated shuttle device. The irradiated samples were available for study and chemical separation within about one second at the end of the irradiation period. When chemical separation of the irradiated sample was required, the chemical separation could be accomplished in a short time after return of the sample. Separation times have ranged from about one second to several minutes.

DISCUSSION

Over the years, various techniques of analyzing the irradiated samples have been employed. Many different chemical separation methods have been developed, used and then abandoned for newer and more sophisticated methods. The shuttle device and the rabbits have been made from a wide variety of

materials, most of which were considered inert. For
the most part, these materials are inert in ordinary
chemical sense of the word, however, under sustained
high neutron and gamma fluxes material such as
"Tygon" and "Teflon" are not inert. Physically, these
materials withstand the radiation reasonably well,
but there is considerable activation of short-lived
materials, namely, fluorine and chlorine.

Rabbits made from "Teflon" became so radioactive that the radiation field from the fluorine (^{20}F)
in the rabbit was frequently greater than the radiation
field from the fission products contained in the rabbit.

Some of the radiochlorine (^{38}Cl) and radiofluorine
is swept out of the tubing and rabbits by the large
amount of gas used to propell the rabbits. In addition,
impurities in the helium were also activated. Some
of these radioactive materials were able to escape
from the shuttle device.

The radioactive material liberated by the best
and most gas tight shuttle system and the chemical
separation operations required continuous air monitoring of the work spaces. For most of these irradiations, the work space was a large trailer (Fig. 1)
outfitted with the equipment necessary to accomplish
the desired results. Work space within this trailer
was restricted (Fig. 2). It was often difficult to find
the space necessary to set up the usual type of air
sampler. In addition, the radiation fields from the
irradiated samples and/or the rabbits prevented the
direct monitoring of the filter medium on the air
sampler.

One of the first methods used to monitor the air
in the working area of the trailer was to sample the
exhaust from the trailer. An exhaust blower with a

Fig. 1. Photograph of Trailer "N-8" and Rad-Safe Van Located at Vallecitos Atomic Products Laboratory Nuclear Test Reactor in Pleasanton, California.

capacity of about 750 CFM was equipped with a MSA
Ultra Filter. A clear plastic "close capture hood"
(Fig. 2), was arranged concentrically with the "rabbit
catcher" end of the shuttle tube. An air sample was
obtained from this exhaust blower at a point just upstream from the filter. This air sample gave some
idea of the concentration of radioactive materials
within the "breathing zone" of the trailer, but unfortunately, the estimate was always high. Usually
it was high enough that we really should have stopped
operating. "Grab" samples taken within the trailer
more nearly represented the true concentration of
radioactive materials, but grab samples were difficult to obtain due to lack of space and the high noise
level of this type of sampler. In fact, the "noise"
problem was the single most objectionable feature of
this type of sampling.

RESULTS

In order to circumvent the problems of lack of
space, high noise levels and inaccurate and often delayed results, it was decided to develop an air
sampler that would operate at a remote location with
little or no attention and give reasonably timely results. Fig. 3 is a photograph of the completed unit.
Fig. 4 shows a block diagram of the unit.

The unit was built into a standard equipment rack
equipped with caster wheels for "portability". The
completed unit weighs about 115 kg. All of the functional parts, except for the "filter tape deck", are
conventional. The pump is a "dry carbon vane" type
designed for essentially continuous operation.

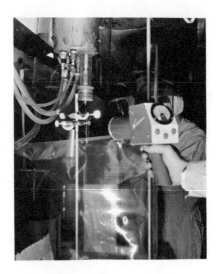

Fig. 2. Photograph of Interior of Trailer "N-8" Illustrating Limited Space and "Close Capture Hood".

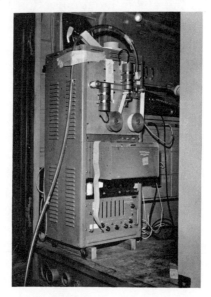

Fig. 3. Photograph of the Completed unit.

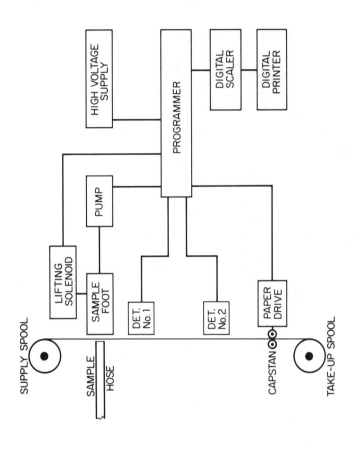

Fig. 4.   Illustrates a Functional Block Diagram of the Unit.

Sampling capability is limited only by the amount of filter tape on the supply spool (and by one's patience in analyzing the resultant data).

The scaler is a Berkeley 7050 with a Berkeley 1452 digital printer. The gating circuits of the scaler have been modified to accept the commands from the programmer. The programmer is a Industrial Timer Corporation, "multi-cam timer" equipped with drive motor and gear combinations to select a time cycle of either 6, 12, 15, or 30 minutes. The programmer is equipped with 7 SPDT switches that provide all of the necessary commands. Other time cycles for the programmer can be selected with a different combination of motors and gears. Fig. 5 illustrates the program used.

The unit will count each incremental sample twice. The first time for 2/6 of a cycle starting shortly after the pump has started. The second count is also for 2/6 of a cycle and begins approximately 5-2/6th cycles after the start of the first count. For the 6 minute cycle, the delay time is 32 minutes.

One inch diameter halogen quenched GM tubes were selected because of their relatively low cost and long life. Replacement has not been necessary but could be accomplished in a few moments.

The GM detectors are shielded with about one inch of lead. This thickness is not enough to reduce the background below about 10-15 counts per minute. In order to assure that the data printed is from the air sample and not from the radiation of either the fission product containing samples or the reactors, the air sampler is usually located at least 30-40 feet from the area where the work is performed.

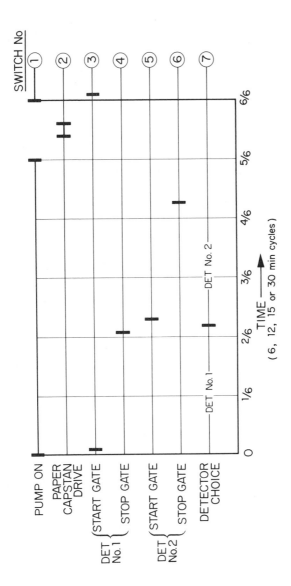

Fig. 5. Sketch Illustrating the Program Used.

The air sample is pulled through a one inch
(nominal diameter) plastic tube. At the measured air
flow rate of about 70.7 liter/minute through CWS-5
filter media there is a lag of only about 6 seconds between the time the air sample enters the plastic tube
and the time of arrival at the sampling head of the air
monitor. At this flow rate with sampling increments
of 5/6th of a 6 minute cycle, and assuming minimum
sample count rate of 10 counts/minute, with a collection efficiency of 83%, the minimum detectable concentration of airborne radioactivity is about 2.6×10^{-10}
μCi/cc (gross beta). This concentration is well within the requirements of 10 CFR 20, especially when an
occupancy factor is used and consideration is also
given to the naturally occurring randon daughters present in the air. In fact, at times the concentration of
randon daughters has exceeded the minimum detection
level by about two orders of magnitude.

In addition to the use of this unit in the irradiation programs for which it was developed, this air
sampler has been used in support of numerous other
projects within the Naval Radiological Defense Laboratory.

While this sampler was developed to fill a specific need, there are certain features that it lacks. Perhaps the most important improvements would be as
follows:

(1) The ability to monitor the sample while it is
being collected. An additional detector could
be mounted such that it monitors the filter
medium in the collection position. This would
require an additional detector and probably a
rate meter recorder combination.

(2) The ability to monitor the ambient radiation with an independent detector. This need was considered but because of the complexity of adding an additional GM tube, this feature could not be incorporated at the time of construction of the unit.

(3) Different filter media. CWS-5 was available in the correct widths. A better media probably would be the glass fiber paper. Unfortunately, this material is not physically rugged enough to be used with this "tape deck".

(4) Alpha monitoring capability is not available. The substitution of scintillation counters equipped with beta and alpha sensitive scintillators would greatly improve the unit. This was considered but could not be accomplished at the time that the unit was built.

SUMMARY

Dut to several problems of lack of space, high noise levels and delayed and often inaccurate results, an incremental air sampler was assembled. The unit has a minimum detection sensitivity of about 2.6×10^{-10} $\mu Ci/cc$. Operation of the unit is completely automatic and the unit can be remotely located. Sampling and digital counting intervals can be changed through the use of other combinations of programmer drive motor and gears. The sampler was developed approximately four years ago and has been used in a number of Laboratory projects.

A COMPARISON OF THE RESPIRATORY DEPOSITION OF AIRBORNE RADIOACTIVITY AS DETERMINED BY GRADED FILTRATION TECHNIQUE AND THE CURRENT ICRP LUNG MODEL

B. Shleien
U. S. Department of Health, Education and Welfare
Consumer Protection and Environmental Health Service
Environmental Control Administration
National Center for Radiological Health
Northeastern Radiological Health Laboratory
Winchester, Massachusetts

ABSTRACT

Data are presented on the particle size of airborne ^{95}Zr-^{95}Nb, ^{131}I, ^{89}Sr, and ^{90}Sr collected at ground-level following two atmospheric nuclear detonations. The particle size and respiratory deposition results obtained by graded filtration technique are compared to ICRP Lung Model data for Activity Median Aerodynamic Diameter (AMAD) versus deposition.

In the pulmonary region, for particle sizes between 0.15 to 3.0 microns in diameter, results are in

1159

agreement. For particles above 3.0 micron, estimated pulmonary deposition ranged from 10 to 17 percent and 3 to 17 percent, respectively, for the ICRP
and the experimental data. Comparisons are also
presented for deposition of airborne radioactive particles in the nasopharyngeal region of the respiratory
tract.

It is concluded that the graded filtration technique
yields results similar to those predicted by the ICRP
Lung Model of Activity Median Aerodynamic Diameter
versus deposition.

INTRODUCTION

Determination of the respiratory tract deposition of airborne radioactive debris is essential for
evaluation of its potential health hazard. Recently,
a method for the determination of airborne particle
size, referred to as "graded filtration-linear programming technique" was proposed.[1] When used with
appropriate computer techniques, this method yields
information on mean particle size, particle size distribution and respiratory tract deposition. It is the
purpose of this paper to present these results as related to airborne fallout following two atmospheric
nuclear detonations. Furthermore, results gained
through use of a "graded filtration-linear programming
technique" for the geometric mean diameter of radioactive fallout versus respiratory deposition are compared to the activity median aerodynamic diameter
(AMAD) versus respiratory deposition proposed by
the ICRP Task Group on Lung Dynamics.[2]

Briefly, a "graded filtration sampler" is a particle size classifying device consisting of a series of three or five filters nested in order of increasing retentivity for particles of decreasing size. The distribution of materials collected on the filters in the nest can be determined by suitable physical or radiochemical analysis. Two models serve to estimate deposition in the respiratory tract.

The first model is based on laboratory calibration of the filters with homogeneous aerosols and relates filter collection efficiency to aerodynamic particle size distribution.[3] The second model relates this distribution to deposition in the nasopharyngeal, tracheobronchial, and pulmonary regions of the respiratory tract based on curves for aerodynamic particle diameter versus deposition (Fig. 4 of Ref. 2). Linear programming computor techniques, using the above models, then relate the whole size distribution of airborne fallout on the graded filters to its deposition in the three regions of the respiratory tract. Modification of the computer technique also permits estimation of the geometric mean (log-mean) particle diameter.

Similar techniques have been employed by other investigators to determine the particle size of airborne radioactivity,[4] and by this author to estimate radiation doses to the respiratory tract from inhalation of airborne radioactivity.[5] This paper presents data on the particle size of airborne fallout following two atmospheric nuclear detonations and compares field and theoretical results for the respiratory tract deposition of airborne fallout.

B. SHLEIEN

METHOD

Results of analysis of airborne nuclear debris from foreign nuclear tests which were reported to have occurred on October 27, 1966 and December 28, 1966, are the basic data used in this paper. Debris from each test reached Winchester, Massachusetts, approximately eight days after detonations. The U.S. Atomic Energy Commission has reported that the first detonation was of low to low intermediate yield (20 to 200 kilotons TNT equivalent), and the second had a yield of few hundred kilotons and both detonations employed enriched uranium,[6,7] Samples of airborne fallout were collected 1 meter above the ground at a sampling rate of approximately 40 cubic feet of air per minute over three day periods. The activities of ^{95}Zr-^{95}Nb and ^{131}I were determined by gamma spectroscope and those of ^{89}Sr and ^{90}Sr by standard radiochemical techniques.[8] The distribution of these radionuclides on the filter nest was used as input data in the computer program and provided results in terms of particle size and respiratory tract deposition estimates.

The "graded filtration-linear programming technique" permits several resultant outputs. They are: (1) A minimum and maximum estimate of the geometric mean diameter of the radioactive particle size distribution. The average of the minimum and maximum estimates is in all cases considered to be the best estimate. (2) A minimum and maximum estimate of deposition in the nasopharyngeal, tracheobronchial and pulmonary regions of the respiratory tract.

RESULTS AND DISCUSSION

The best estimates of the geometric mean diameter for specific radionuclides in selected samples are presented in Table I. Differences in particle size between specific radionuclides and samples are indicated. These data are representative of a larger number of samples analysed for which results are plotted in this presentation later.

The ICRP Lung Model relates percent deposition in the three regions of the respiratory tract to the Activity Median Aerodynamic Diameter (AMAD) of an aerosol (Figs. 13 and 14 of Ref. 2). Initially, it appeared that deposition curves might be required for a wide variety of geometric standard deviations.[2] However, when the deposition curves were applied to hypothetical aerosols having log-normal activity distributions, it was found that the predicted deposition in each of three compartments could be related to the median diameter of the distribution almost independently of the geometric standard deviation. Since the graded filtration particle size classifying device is calibrated in terms of aerodynamic diameter, it can be assumed that the geometric mean diameter of the radioactive debris and the AMAD are essentially equivalent, if the particle size distribution of the radionuclide is log-normal. Although this may not always be true, the fact that respiratory tract deposition in the ICRP Lung Model is almost independent of the geometric standard deviation would appear to negate differences due to non-homogeneity of an aerosol as it influences deposition.

A comparison of the estimates of the geometric mean diameter versus deposition obtained by graded

Table I.

Particle Size

Sample Designation	Collection Date (Inclusive)	Geometric Mean Diameter (Microns)
Sample A	Nov. 3, to	
	Nov. 6, 1966	
	^{95}Zr-^{95}Nb	6.2
	^{131}I	0.6
	^{89}Sr	1.6
	^{90}Sr	1.0
Sample B	Nov. 24 to	
	Nov. 27, 1966	
	^{95}Zr-^{95}Nb	1.6
	^{131}I	0.3
	^{89}Sr	0.2
Sample C	Jan. 4, to	
	Jan. 7, 1967	
	^{95}Zr-^{95}Nb	7.5
	^{131}I	2.1
	^{90}Sr	1.0
Sample D	Jan. 7, to	
	Jan. 10, 1967	
	^{95}Zr-^{95}Nb	1.5
	^{131}I	0.1
	^{89}Sr	0.4
	^{90}Sr	0.5
Sample E	Jan. 19, to	
	Jan. 22, 1967	
	^{95}Zr-^{95}Nb	4.9
	^{131}I	1.2

filtration-linear programming techniques was made
to the ICRP curves of AMAD's versus deposition.
Comparisons for deposition are presented in Fig. 1
for the pulmonary region and in Fig. 2 for the
nasopharyngeal region. The solid lines in these
figures are re-drawn from the Task Group Report
(Figs. 13 and 14 of Ref. 2). Those results for which
data are presented in Table I are circled and designated by the appropriate alphabetic letter in those
figures. Deposition in the tracheobronchial region
based on AMAD's is constant and thus a comparison
for this region is not made.

In the pulmonary region, for particle sizes between 0.15 and 3.0 microns in diameter results of
field testing and the ICRP Model are in agreement.
For particles above 3.0 micron, estimated pulmonary
deposition ranged from 10 to 17 percent and 3 to 17
percent, respectively, for the ICRP curves of AMAD
versus deposition and the experimental data. The
reason for the disagreement is that the deposition
parameters used in the graded filtration-linear programming were obtained from the ICRP curve for
deposition as a function of aerodynamic particle size
for 15 respirations per minute, with 1450 ml as the
tidal volume (Fig. 4. of Ref. 2). Above 5 micron
diameter these parameter values are lower than those
used by the ICRP for AMAD versus pulmonary deposition.

The comparison of results for deposition in the
nasopharyngeal region (Fig. 2) indicates a consistent
elevation in estimated deposition when the field results are compared to AMAD versus deposition. This
difference is 12 percent, 20 percent, and 10 percent
maximum at particle diameters of 0.3, 1.0, and 5.0

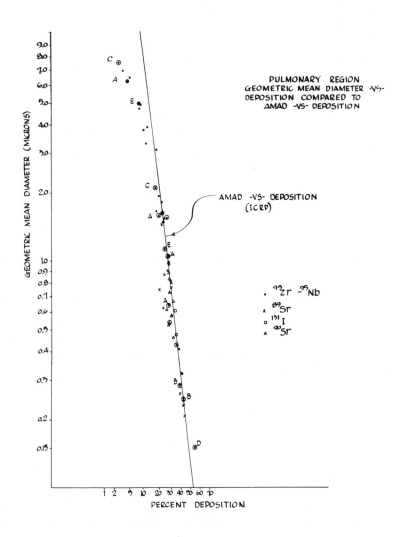

Figure 1.

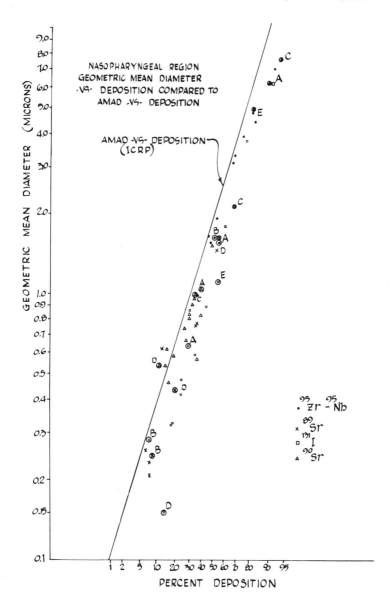

Figure 2.

B. SHLEIEN

microns, respectively. In contrast to the case for the pulmonary region, there are no differences in the deposition parameters from the two sources. Therefore, a systematic difference in the graded filtration-linear programming technique for the nasopharyngeal region appears to be operative.

The constraints on the computer program are such that the estimates of minimum and maximum deposition in the pulmonary region have a lesser range than estimates for the pulmonary region, therefore, are closer to the AMAD versus deposition curve than those for the nasopharyngeal region.

SUMMARY AND CONCLUSIONS

The graded filtration-linear programming technique can be used to obtain: (1) the geometric mean particle diameter of an airborne radioactive aerosol; and (2) an estimate of regional respiratory deposition under field conditions. Comparison of results obtained by this technique and the ICRP curve for AMAD versus deposition indicates good agreement in the pulmonary region. Comparison of results in the tracheobronchial are not made because a constant deposition value is used for AMAD's versus deposition in this region. In the nasopharyngeal region, the experimental technique introduces a systematic difference which gives deposition results between 10 and 20 percent greater than the curve for AMAD versus deposition.

REFERENCES

1. B. Shleien, A. G. Friend and H. A. Thomas, Jr., Health Physics, 13 : 513, (1967).

2. Task Group on Lung Dynamics (Committee II of the International Commission on Radiological Protection), Health Physics 12 : 173, (1966).

3. B. Shleien, J. A. Cochran and A. G. Friend, Amer, Ind, Hyg. Assoc. J., 27.353, (1966).

4. L. B. Lockhart, Jr., R. L. Patterson, Jr., and W. A. Saunders, Jr., J. Geophys. Res., 70:6033, (1965).

5. B. Shleien, M. A. Wall and D. Lutz, Envir. Sci. Tech. 2:438, (1968).

6. U. S. Atomic Energy Commission, News Releases of October 28 and November 2, 1966.

7. U. S. Atomic Energy Commission, News Releases of December 28, 1966 and January 3, 1967.

8. P. Magno, E. Baratta and E. Ferri, "Analysis of Environmental Samples—Chemical and Radio-chemical Procedures", NERHL Report 64-1, Northeastern Radiological Health Laboratory, Winchester, Massachusetts, (1964).

RADIOACTIVITY AIR MONITORING
IN MEXICO

M. B. Skertchly, R. M. Nulman and M. B. Vasquez

Comision Nacional De Energia Nuclear
Piso, Mexico

In 1956 in answer to the requirements of the
U. N. Committee for the study of the effects of Ionizing Radiation on Man, the Nuclear Energy Commission of Mexico contracted the Institute of Physics of
the University of Mexico to undertake the study of
fallout from nuclear debris.

To accomplish this study five sampling stations
were set up in different parts of Mexico. Samples
were obtained using gummed paper and water pan
techniques. Only the gross beta activity from these
samples was measured.

In 1961, the Nuclear Energy Commission of
Mexico instructed its Radiological Security Division
to undertake this work. Sixteen stations were established throughout the Mexican Republic. Samples of
air particulates were obtained using glass fiber filter

papers. After a statistical study of the results, the number of stations was reduced to 14 in 1966 because large differences between nearby stations were not observed and because the levels had decreased.

The stations are operated by various personnel, depending on their locations. At airports, airline personnel are used. Other stations are operated by personnel from the Centro de Prevision del Golfo de México, the Department of Chemistry of the University of Mérida, and the Superior School of Maritime Sciences of the University of Baja California.

Until 1966 only the gross beta activity of all samples was measured. Since 1967 radiochemical analyses for ^{90}Sr and ^{137}Cs have been performed on samples from Chihuahua, México City, and Mérida on a bimonthly basis. These analyses are performed on samples which indicate air contamination in excess of 1 pCi/m^3 of gross beta activity in order to determine 1/10 MPC values applicable for the general population.

At all but three stations we considered it sufficient to measure only the gross beta activity of the air samples because the general radioactivity level in Mexico is very low. Most of the activity results from a slow fallout of materials stored in the stratosphere. This process has been observed by meteorological studies described later in this paper. Only one research reactor exists in the country and hence no local source of contamination is expected.

At three stations located in the Northeast, center, and southwest parts of Mexico, samples are analyzed for ^{90}Sr and ^{137}Cs when the gross beta activity for the month shows an average value higher than 1 pCi/m^3.

These analyses are performed in order to provide a general idea of the amount of beta emitters with high radiotoxicity.

LOCATION OF SAMPLING STATIONS

In order to locate sampling stations equally spaced throughout Mexico, equalateral triangles were layed out on a map of Mexico. Each triangle side was the equivalent of 350 km and stations were located near each triangle vertex at a site which had regular airline communication with Mexico City and electric power. The topography of the location was also taken into account since the measured activities at each station should be representative of the average activity of the area. Stations were not installed in deep valleys, sites surrounded by mountains nor on hill tops. Figure 1 shows the geographical location of the sampling stations and Table 1 summarizes their latitude, longitude and altitude.

SAMPLING

Sampling is performed with "Staplex" high volume samplers model AF 315 operating at 1765 rpm. They are usually mounted on the roof of the highest building in the neighborhood; this places them about 6 or 7 meters above the ground. Valves and filtration tubes are oriented upward to permit a uniform air flow and the equipment, including a current regulator, rotameter and stopclock, is placed in a housing similar to the ones used for meteorological equipment. Samplers are used with a 6" x 9" adapter

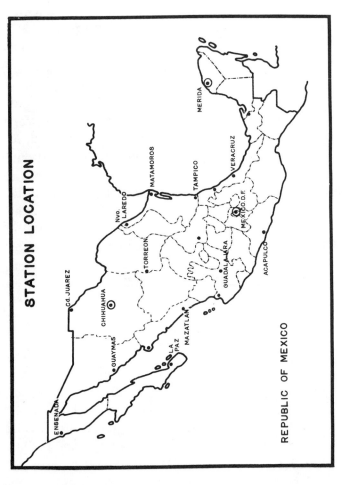

Figure 1.

TABLE No. 1

STATION	LATITUDE	LONGITUDE	ALTITUDE
ACAPULCO, GRO.	16⁸ 50' 21"	99⁰ 55' 01"	2 Mts.
CD. JUAREZ, CHIH.	31⁰ 44' 19"	106⁰ 29' 15"	1,144 Mts.
CHIHUAHUA, CHIH.	28⁰ 38' 12"	106⁰ 04' 42"	1,430 Mts.
ENSENADA, B.C.	31⁰ 51' 19"	116⁰ 38' 09"	2 Mts.
G UADALAJARA, JAL.	28⁰ 40' 32"	103⁰ 23' 09"	1,567 Mts.
LA PAZ, B.C.	24⁰ 09' 41"	110⁰ 24' 44"	10 Mts.
MATAMOROS, TAMPS.	25⁰ 52' 45"	97⁰ 31' 09"	12 Mts.
MERIDA, YUC.	20⁰ 59'	89⁰ 38' 53"	9 Mts.
MEXICO, D. F.	19⁰ 25' 59"	99⁰ 07' 58"	2,300 Mts.
MAZATLAN, SIN.	230⁰ 18'	106⁰ 24'	2 Mts.
NUEVO LAREDO, TAMPS.	27⁰ 29' 48"	99⁰ 30' 01"	171 Mts.
TAMPICO, TAMPS.	20⁰ 13'	97⁰ 51' 19"	12 Mts.
TORREON, COAH.	25⁰ 32' 18"	103⁰ 27' 55"	1,140 Mts.
VERACRUZ, VER.	19⁰ 12' 02"	96⁰ 08' 13"	14 Mts.

in which glass fiber filters ("Staplex" TFA 69, 15.24 x 22.86 cm.) are fitted. These filters have an effective area of 260 cm^2. The vacuum used is 3.2 mmHg and the flow through the equipment is 1.56 m m^3/min.

The sample volume is determined by averaging the initial and final flow as measured by the rotameter. Multiplying the flow per hour by the total number of hours we get the total volume of air. This is between 1000 and 2000 m^3 per day depending on the altitude of the sampling station.

GROSS BETA ACTIVITY

The gross beta activity is measured by counting a circular filter paper sample in a Nuclear Chicago low background anticoincidence counter. The samples are 2.5 cm in diameter and are cut at random from the filter paper. The counter background is 2.5 cpm and its minimum detectable activity is 0.1 pCi/m^3 ± 0.5 pCi/m^3 at the 95% confidence level. The counter is standardized with ^{204}Tl and the results are corrected for self absorption due to dust collected on the filter paper.

We chose ^{204}Tl as a standardizing source because the counter's response to betas from this isotope is similar, but slightly lower, than the response to fission product betas. Thus the measured efficiency is lower than the real efficiency for fission products and the calculated fission product activity higher than the actual activity.

Filters are humidified to a constand humidity by allowing them to stand in a high humidity chamber for several days. They are then weighed and sent to the sampling stations. When the filters arrive back at our laboratory they are placed in the same humidity chamber for three days. This allows the activities due to radon and thoron to decay and the humidity to return to its original value. They are then weighed mounted on aluminum planchets, and counted in the low background anticoincidence counter.

The calculation of the activity is made by the following formula:

$$pCi/m^3 = \frac{c.p.m. \ (St)}{(E) \ (2.22) \ (T) \ (Sc) \ (V)}$$

when: cpm = net counts per minute

St   = total surface of the filter paper

Sc   = planchet surface

E   = counter efficiency for ^{204}Tl on aluminum planchets

T   = transmission factor

V   = volume of air in m^3

^{90}SR AND CS^{137} MEASUREMENT

^{90}Sr and ^{137}Cs are measured pooling together and processing 2 months' filters (15-30 filters). Filters are refluxed in a 2 liter beaker with 5 ml. of HC1 per filter and enough water to cover the filters. After one and a half hours of reflux, they are allowed to cool and are filtered.   The filtrate is evaporated to dryness and 1 ml. each of Cesium carrier (10 mg/ml) and strontium carrier (88 mg/ml) are added with concentrated nitric acid to dissolve the residue.

The sample is evaporated to a white residue which is redissolved in as little nitric acid as possible and diluted with water to 1 liter in order to obtain a solution with pH2.   This solution is used for the determination of ^{137}Cs and ^{90}Sr.

CESIUM 137

Ammonium phosphomolibdate (AMP) is used in the chemical preparation of the ^{137}Cs samples.   The

AMP is deposited on a filter paper to a thickness of 20 to 30 mg/cm^2. The AMP is suspended in 1% nitric acid, filtered through the millipore filter to obtain a uniform layer and washed with a little more 1% nitric acid, taking care not to remove the AMP layer. The sample is filtered through this layer at a maximum flow of 20 ml/min. per cm^2 of filter surface. ^{137}Cs is absorbed on the surface of the AMP layer, and no self-absorption factor is necessary. The filtrate is used to measure the ^{90}Sr activity. The millipore filter with the AMP precipitate is mounted on a planchet for counting.

The chemical yield may be obtained by tagging the filtrate with a known amount of ^{137}Cs, repeating the procedure described above and calculating the ^{137}Cs retained on the AMP layer.

Calculations:

The activity is calculated as follows:

$$a = \frac{n}{2.22\,N} \pm Q$$

and N = ACE

where   a = activity in pCi/m^3

          n = net counts per minute

          A = counting efficiency including backscatter factors

          C = chemical yield

$$E = \text{sample volume in m}^3$$

$$Q = \text{statistical error at a 95\% confidence level}$$

Q may be calculated by:

$$Q = \frac{1.96 \sqrt{e+b}}{2.22 N}$$

where: b = counts per minute of background
                divided by the background counting
                time in minutes

                c = counts per minute of sample divided
                by the sample counting time in min-
                utes.

STRONTIUM 90

The filtrate passed through the AMP layer during the ^{137}Cs determination is evaporated to 250 ml.
and 25 g. of oxalic acid are added. The solution is
warmed to 80° C and a solution of 1:1 of ammonium
hydroxide is added with stirring until pH5 is obtained. The oxalates are left over night, filtered
through Whatman No. 42 filter paper and the filter
paper with the precipitate is incinerated in a muffle
furnace (650°C for 4 hours).

The crucible with the precipitate is cooled to
room temperature. Ten drops of concentrated nitric
acid and 2 ml. of H_2O_2 are added and the slurry is

dryed on a hotplate. The residue is dissolved in 50 ml. of 0.08 N HC1 and the solution is extracted twice with 50 ml. of 20 % HDEHP (Di-2-ethyl-hexil-phosphoric acid) in toluene. After the second extraction, time is noted. The aqueous phase is then stored for 2 weeks for yttrium growth.

After 2 weeks, the ^{90}Y is extracted by 10% HDEHP in toluene. The aqueous phase is used for chemical yield determination, and the organic phase is back-extracted twice in 3 N HNO$_3$. The nitric solution is evaporated and transferred into a stainless steel planchet, to be counted. Time is noted.

Calculations:

$$a = \frac{n}{2.22\,N} \pm Q$$

$$N = ABCDEF$$

$$Q = \pm\ \frac{1.96\,\sqrt{e+b}}{2.22\,N}$$

Where

$a = {}^{90}Sr$ activity pCi/m^3

n = Net count (c.p.m.)

A = Counting efficiency including backscatter factor.

B = Growth factor for ^{90}Y, after two weeks
$B = 0.97$

C = Extraction efficiency (0.97)

D = Decay factor for ^{90}Y (from extraction to counting)

E = Volume of air filtered in m^3.

F = Chemical yield

b = Counts per minute of background divided by counting time.

e = Counts per minute of sample divided by counting time.

METEOROLOGICAL STUDIES

In order to find out if there was a pattern relating meteorological data with radioactivity in air particulates, daily diagrams of maximum and minimum temperatures, relative humidity and atmospheric pressure were compared with daily radioactivity data from one station; the following was observed: When the relative humidity and maximum and minimum temperatures diminish, an increase in activity is noted. Also a sudden increase in atmospheric pressure correlates with an increase in activity. From these comparisons of surface data it was concluded that an advection of dry cold air, with an increase in pressure, increases activity.

From studies of the movements of air masses, it was observed that the displacements of polar fronts are associated with increases in activity. Surface winds were also related to activity since an

increase in activity was observed where convergent
vectors came from the north and activity was lower
in zones where convergent vectors came from the
South.

The changes in the upper atmosphere were also taken into account by studying the relationship
between radioactivity in fallout at the 400 mbs and
200 mbs levels. At out latitudes the 400 mbs level
is in the troposphere and the 200 mbs level corresponds to the tropopause. It was observed that
when these constant pressure levels are near the
earth's surface, activity increases and when they
move away less radioactivity is observed.

These changes in the altitude of atmospheric
constant pressure levels are related with vertical
movements of air molecules and suspended particles
in it. Since in Mexico the main source of radioactivity is now expected to come from the stratosphere,
a downward movement of air and particles from the
stratosphere, together with an increase in atmospheric pressure and a decrease in temperature,
produces higher radioactivity levels on the earth's
surface.

From studies of the jet stream we could also
conclude that it injects radioactive particles from
the stratosphere. Where the jet stream is nearer to
the earth's surface, vertical windshear and turbu-
lence produce increasing radioactivity near the surface.

Monthly isograms are made using the average
radioactivity data and the meteorological factors as
mentioned above. The most affected areas of Mexico

can be delineated by these studies even with the sampling stations far away from each other. The affected areas can then be controlled in case of dangerous increase in activity.

SUMMARY

This paper presents the air monitoring program followed in Mexico. It was designed to fit the following conditions prevailing in our country:

1. Geographical and meteorological conditions have prevented our country from considerable contamination from nuclear debris. Almost all activities being detected are falling to the earth's surface from the stratosphere where they were previously injected. ^{89}Sr and ^{131}I are not being measured.

2. There is no point source of contamination, since no large reactors nor fuel reprocessing plants are in operation in Mexico. This enables us to use a triangular pattern for station location.

3. Sampling must detect any dangerous increase in activity in any part of our territory, and give a general idea of the exposure of the population, and yet it must be inexpensive.

To satisfy these conditions, we make isograms of gross beta activity. Radiochemical analysis are only made when gross beta activities are higher than 1/10 of M. P. C. Since we have to use personnel who are not working for us and are not trained in operating counting equipment, samples are returned for counting.

Our present program ahs fulfilled these requirements.

ACKNOWLEDGEMENT

We are indebted to Eng. Widmar for his valuable comments and discussion of the manuscript.

BIBLIOGRAPHY

Primero, segundo, tercero, cuarto y quinto informe sobre Estudios de la Precipitación Radiactiva en México. - Revista Mexica na de Física 5: 153-166 (1956); 6: 97-104 (1957); 8: 61-85 (1959); 10: (1961) y C. N. E. N. : 1968.

Vásquez M. B. , De Nulman R. M. and Skertchly M. B. "Informe para el Comité Científico de las Naciones Unidas para el Estudio de los Efectos de las Radiaciones Ionizantes, Analisis Radioquimicos en Muestras Ambientales en México durante 1966". Comisión Nacional de Energía Nuclear. 1967.

Burillo G. "Análisis Químico y Radioquímico de la Precipitación Atmosférica". U. N. A. M. Facultad de Ciencias Quimicas. 1966.

Session VII
EXPOSURE AND CONTAMINATION CONTROL

Chairman

L. J. BEAUFAIT
U.S. Atomic Energy Commission
Berkeley, Calif.

C. MEINHOLD
Brookhaven National Laboratory
Upton, N.Y.

PRACTICAL ASPECTS OF
EXPOSURE CONTROL

Victor M. Milligan
Reynolds Electric & Engineering Co., Inc.
Las Vegas, Nev.

When attempting to organize material on "Radiation Exposure Control," it soon becomes obvious that the time allotted must be spent in discussing some general (but hopefully significant) facets of the subject. My intent is to use the available time to discuss what I consider to be some practical aspects and operational considerations, recognizing that other factors strongly relate to the problem.

Although I have elected to discuss Exposure Control primarily from the standpoint of the applied health physicist working "in the field," I wish to emphasize that I consider the health physics engineering that goes into a facility, or piece of equipment, or an experiment, to be the primary and probably most important phase of exposure control. Here it becomes possible to "engineer out" the radiation hazard by "designing in" the basic principles of radiation protection. Here is where the principles of shielding, distance, containment, and ventilation control may be implemented. This truth has been demonstrated time

1187

and again in my own experience, and its importance
cannot be diminished. Nevertheless, the specific
problems of the operational health physicist are
real and challenging and certainly worthy of recogni-
tion.

Health physicists tend to be specialized, and this
is consistent with the way our technological society
operates. They may work in nuclear reactors (a
variety of types with each having unique characteristics), in chemical processing plants, research laboratories, accelerators, medical facilities, or nuclear
testing facilities, to mention some of the major cate-
gories. Wherever he works, it is important that he
be intimately familiar with the operation which he is
supporting, if he is to contribute to the goals and objectives of his organization while accomplishing his
own professional objectives. No matter what the degree of his specialization, one of the common denomi-
nators of responsibility will be that of the measurement and control of radiation dose to personnel.

It is appropriate here to talk briefly to the matter
of exposure limits as viewed by the operational health
physicist. Actually, this individual in his day-to-day
activity is relatively unconcerned with how the limits
are established. In this respect, he might be considered in the same light as the practicing attorney
who operates within a "closed system" of established
rules and laws, not concerning himself with how or
when the laws are legislated. Of course, from an
academic or detached viewpoint, the health physicist
probably has studied the history and rationale of
radiation limits and has considerable background and
knowledge thereof. But whether the limits are
established by national law, an atomic energy agency,

state law, or military regulation, he is primarily concerned with how to operate within the established system in an effective and economical way.

A very important consideration in the establishment of an adequate exposure control program has to do with the health physicist and his ability to work with management to establish work procedures, operational guides, and administrative rules. The excellent radiation safety record in the nuclear industry today must be closely related to a general mutual understanding and respect between management and the health physicist.

Then, if exposure limits are already established, and strict administrative controls which limit personnel dose to a small amount daily can be enforced, it seems that there should be no problem whatsoever in performing operations within acceptable limits.

Obviously, this logic has serious faults. If an operation can be performed with the minimum number of required personnel, and if radiation dose rates are such that low daily dose limits can be easily met, then the expenditures for an exposure control and personnel dosimetry program can probably be minimal. On the other hand, if dose rates are high, we may find that arbitrarily low daily dose limits may be exceeded within a few hours of work per day. If so, additional personnel will be needed to perform the operation, and labor costs will rise. Only a few qualified specialist personnel may be available, and if they are not allowed to perform their functions, the productive work of the operation may not be accomplished. Existing labor contracts may require premium wages for shortened work days and uneven distribution of overtime among various crafts. With this basis for disagreement, union demands for preferential pay may be increased.

When exposure rates are high enough to cause concern about personnel doses near permissible limits, the temptation is to implement arbitrarily low administrative limits. Here lies the challenge to the operationally-oriented health physicist; to utilize his technical knowledge and common sense toward implementation of time limitation techniques that will assure full use of available exposure and still remain within acceptable limits.

The use of strict and arbitrary administrative controls may seem to be an easy, dignified approach which demonstrates the influence and stature of the health physics establishment, but this approach may prove costly. Some organizations in the nuclear industry have undergone difficult labor negotiations because instead of considering radiation exposure as a routine operational problem they overemphasized the hazard by using arbitrarily low dose limits. In one case, dose limits more restrictive than required by AEC regulations have become part of a union contract, and, on the national level, automatic compensation for amount of dose received has actually been considered.

1. EXPOSURE CONTROL CONSIDERATIONS

The United States Atomic Energy Commission has established an occupational exposure limit of 3000 mrem per quarter year for whole-body, penetrating radiation dose. Using this as an example, the health physicist for a particular AEC facility is concerned with accomplishing the operation without exposing personnel to even a few mrem above the limit. In so doing, he is not particularly concerned with whether a 3000 mrem dose in a quarter year is harmful; neither is he concerned with whether a 1000 or 2000 mrem dose per quarter is harmful. In fact, his only concern above performing the operation within the limit is assuring that personnel are not exposed unnecessarily. This concern is both ethical and economical, and it does not mean that he should have the prerogative to increase operational costs simply to keep personnel doses as low as possible. A major problem that faces the health physicist is the possible error between field dose estimates and recorded results from personnel dosimeters. The possible error can be increased by the time lag between receiving dosimeter results and additional exposure, and inadvertent or unplanned exposure after most of a limit has been accumulated adds to the problem. Many installations allow for those errors by using exposure guides that are below the established limits. For example, a 2500 mrem per quarter guide might be enforced in place of the 3000 mrem limit previously mentioned. From an economic standpoint this practice could be questioned under certain circumstances; however, I know of no better answer to avoid the occasional error which would throw a personnel dose into the incident category.

It is sometimes advantageous to prorate dose during critical operation to ensure the availability of specialist personnel throughout the operation. In such cases, the necessity for proration should be mutually agreed upon by operational supervision and the responsible health physicist, and every effort should be made to assure that personnel so restricted can complete their normal work days in non-exposure areas. The point again emphasized is that radiation dose should not be prorated just to keep personnel doses as low as possible on a daily or weekly basis. This practice can result in a multitude of personnel, scheduling, and labor problems which may seriously hamper and/or increase the costs of an operation. Additionally, operating personnel may become frustrated by restrictions which prevent them from accomplishing their objectives in a timely and effective way, and work performance may noticibly deteriorate. Thus, arbitrarily low administrative dose limits for operating personnel can cause far-reaching operational problems. Those problems can best be avoided by working closely with operational supervision and carefully considering their objectives, resources, and responsibilities. Such consideration is usually rewarded with full cooperation toward successful implementation of realistic exposure control procedures.

2. EXPOSURE CONTROL PROCEDURES

Some of the basic tools for controlling accumulation of dose from external sources are on-the-spot measurements, self-reading pocket dosimeters, personnel dosimeters, dose rate information, and records of radiation area entries. Pocket dosimeters, dose rate information, and records of previous entries can be used to estimate accumulated dose over relatively short periods of time, but, particularly in high dose situations, the possible errors involved dictate that the periods of time covered by estimates be kept as short as practical.

The fast-moving technology of thermoluminescent dosimetry offers the potential for eventual quick field determination of personnel exposures, and is already useful in the determination of extremity exposure. Film finger rings and wrist badges have built-in limitations but TLD devices seem to lend themselves to this type of measurement. As TLD readers become more portable and less expensive, these devices should prove even more useful in rapidly determining individual exposures.

Dose reports are prepared for two reasons, The first and most obvious reason is to demonstrate that dose to personnel has been kept within established limits. These dose reports are usually prepared and transmitted to various organizations weekly, monthly, or quarterly—just for the record. How current these records are is only important from an administrative viewpoint.

The second and perhaps most important reason for dose reports is exposure control. Health physics operational personnel must be kept as current as possible on the amounts of dose accumulated by persons who are receiving more dose each day.

In some nuclear operations, it is not uncommon for certain personnel to receive their acceptable quarterly dose in two weeks, a week, or even a few days. In such cases, it is helpful to the health physicist to have a daily dose report for all personnel who have received a reportable dose for the year or quarter. These official reports can thus be checked and used as a control instrument each time a person enters a controlled radiation exposure area. They are also invaluable for preplanning where repetitive exposure situations occur.

Here, a word might be said relative to the accuracy of the results of several dosimeters changed throughout an exposure period as opposed to a single dosimeter worn over the same period. Our experience indicates that little or no sacrifice of accuracy occurs with the former practice if appropriate limits for exchange are prescribed.

3. EXPOSURE CONTROL INTERNAL SOURCES

Personnel exposures resulting from internally deposited radionuclides represent perhaps the thorniest dose-control problem. Much of the applied radiation protection effort in this area is directed towards after-the-fact documentation of such exposure. Air sampling, urinary bioassay sampling, and organ scanning with gamma spectrum devices all contribute useful information relative to de facto exposures, but are too often in-applicable to future internal exposure control except in those cases where work is routine or repetitive in nature. Too often, an ultraconservative approach is adapted which dictates that we assume that the worst exposure conditions are likely to occur and therefore require the workers to wear the maximum amount of protective clothing and respiratory equipment. Therefore, the best available methods of internal exposure control are rigorous pre-planning and experience. Avoidance of "over-control" requires the exercise of sound judgement in determining beforehand the probable risks. Considerable thought must be given to the use of respiratory protection equipment in work areas that are physically hazardous. At some time in the course of his professional

employment, the health physicist may find himself
facing a situation whereby some exposure is prefer-
able to compounding an already hazardous situation.
Such circumstances might involve underground work
or work at heights where limited vision or agility
could result in a fatal accident.

One of the most commonly used methods for so-
called "internal exposure control" is urinary sampling.
It bears repeating here that urine sampling can only
tell what has already occurred and therefore, the key
to its use depends upon the applicability of the data
to the next job. Too often, the basis for a bioassay
program seems to be that the more samples the better,
as if there were safety in numbers alone. In fact,
too frequent sampling often results in a reduction of
cooperation from those sampled. Perhaps more consideration should be given to (1) when to obtain a
sample, (2) what to analyze it for, and (3) interpretation of results. It should also be pointed out that it
is all too easy to develop a false sense of security
from bioassay results. The absence of positive results does not necessarily imply that no deposition
occurred. Here again, let me emphasize that pre-
planning and experience as well as a thorough knowledge of the inherent limitations of bioassay techniques are the best safeguards. For some nuclides
such as tritium, grab samples a few hours after
exposure are perfectly adequate. For general fission
product sampling, simulated 24-hour sampling is
acceptable since this has the advantage of not requiring personnel to carry sampling equipment with
them all the time. In the case of radioiodine, urine
sampling is questionable. At best, radioiodine bioassay can be used only as an indication of intake with

organ-scanning techniques as the final basis for dose assessment.

SUMMARY

In summary, the obvious exposure control consideration is the amount of radiation dose allowed under established limits, but the health physicist who implements exposure control procedures faces other problems, the solutions to which may unjustifiably increase operational costs, and prevent completion of a particular operation. Workable solutions to these problems seldom originate at the desk of an administrator, but are found by the consciencious health physicist who works closely and cooperatively with operational personnel.

Dose reports are a valuable tool for control of accumulated dose—if used properly. The maximum benefit insofar as dose control is concerned centers around the problems of data retrieval. Warehouses full of dusty and unused records, except in the relatively rare instances of law suits years after the fact, are useless to the operational health physicist. More effort, therefore, might be expended to extract from exposure records statistical information relative to exposure trends, frequency of occurrence of particular radionuclides, etc., to aid the health physicist in pinpointing potential problem areas and in pre-planning.

A maximum of available capability and funding should be expended toward utilizing dose reports to effectively control exposure, and subsequently decreasing operational problems and costs.

Efforts and expenditures toward creating extensive and redundant record systems, both at the organizational and national levels, should be questioned from the standpoint of their contribution to the real exposure control programs. Solutions to the real exposure control problems include keeping the Health Physicist current on doses accumulated by personnel involved in his operation, avoiding unnecessary or unrealistic dose limits and control procedures, and cooperating with operational personnel to achieve mutually agreed upon objectives.

Finally, a successful exposure control program, whether external or internal in scope, starts with rigorous pre-planning, facility engineering and design, where basic principles of radiation protection may be implemented, followed by the efforts of the field health physicist in dealing with day-to-day problems in an effective and economical manner.

CONTROL AND MEASUREMENT
OF HAND EXPOSURE*

G. D. Carpenter and W. P. Howell
Battelle Memorial Institute
Pacific Northwest Laboratory
Richland, Washington

ABSTRACT

Because limits for extremity dose are much
larger than those for whole body dose, it is sometimes
assumed that extremity dose control does not require
much attention. This may be true where exposures
are entirely in relatively uniform fields of radiation,
but where the extremities may be exposed to much
greater levels of radiation than the whole body, a
program of control and measurement of these doses
is required comparable to that used in control and

*This paper is based on work performed under
United States Atomic Energy Commission Contract
AT(45-1)-1830.

measurement of doses to other organs. The hands
are the extremities most often exposed to radiation
levels not reflected in the results of dosimeters worn
for measurement of whole body dose. In many cases,
the hands may be in direct contact with radioactive
materials for significant periods of time. Because of
inherent difficulties in obtaining accurate contact
measurements on such materials, the health physicist
is sometimes required to make special evaluations
of hand exposures for a given work project. This
paper describes the basic methods and techniques
used in controlling and measuring hand exposure in a
large laboratory.

INTRODUCTION

Whole body radiation dose has traditionally received more attention than localized doses to the extremities. In view of the greater potential for serious
injury when the whole body is exposed, this emphasis
has been proper. However, health physicists sometimes need to be reminded that the first radiation
injuries were to the hands and that the hands continue
to be the most common site of radiation injury.[1]

PERSONNEL EXPOSURE LIMITS AND CONTROLS

Over the past 20 years, responsible technical
bodies in the field of radiation protection have carefully defined the relative importance of radiation dose
to various body organs, including the hands. Their
recommendations appear in the form of occupational

dose limits in the regulations of the Federal Government and the regulations of the agreement states. Contractors to the Atomic Energy Commission such as Pacific Northwest Laboratory are required to control occupational personnel dose within the limits of AEC Manual Chapter 0524.   The limits given in this Manual Chapter are as follows:

Occupational Dose Limits

	Accumulated	Annual	Calendar Quarter
Whole Body, Gonads, Bloodforming Organ	5(N-18)rems*	12 rems	3 rems
Specific Organ other than those listed (e.g., Lung, GI Tract, Lens of the eye).	—	15 rems	5 rems
Bone, Thyroid, Skin of the Whole Body	—	30 rems	10 rems
Skin of the Hands, Forearms, Feet and Ankles	—	75 rems	25 rems

*Where N is the age in years, and is greater than 18.

From the point of view of the operational health physicist, this table of limits poses several problems. First of all, where the exposure to radiation involves an external source, it is impractical to limit the dose to a given internal organ to one value and the dose to another immediately adjacent organ to another value. In a practical sense, the control system must utilize the lowest organ limit in such cases. Second, where personnel are to be exposed to radiation throughout the year, dose fractionation is necessary over periods shorter than a calendar quarter. Third, existing systems for measurement of radiation dose contain inherent errors and have important limitations in measuring dose under field conditions. To provide the operational health physicist with a more workable system, and partial answers to these problems, Pacific Northwest Laboratory personnel have developed the following set of Operational Controls:[2]

Operational Controls

	Annual	Monthly
Dose to the Whole Body	4 rems	1 rems
Dose to the Skin of the Whole Body	24 rems	3 rems
Dose to the Extremities	60 rems	8 rems

Comparison of these values with the occupational limits in the AEC Manual will show that personnel dose maintained within these Operational Controls will be maintained within AEC requirements. The conservatism in the annual Operational Controls provides for the possibility of localized exposures not recorded on dosimeters, discrepancies in dosimeter performance, and situations involving exposure to personnel not wearing dosimeters.

For routine personnel dose control, the Operational Controls are utilized as if they were real limits. When appropriate, personnel are permitted to exceed an Operational Control, but only after careful evaluation of potential errors in measurement of dose already received and in the planned dose to be received. Laboratory accidents occasionally result in personnel dose exceeding the Operational Controls, and sometimes the Occupational Dose Limits. Such situations are carefully evaluated to determine the true radiation dose status of the involved personnel. Only in serious emergencies are personnel authorized to exceed the Occupational Dose Limits.

DETERMINATION OF RELATIVE
IMPORTANCE OF DOSE RATE

Routine personnel dose control at Pacific Northwest Laboratory utilizes the monthly Operational Controls in the following manner. Each situation involving exposure to radiation is evaluated by Radiation Monitoring (health physics) personnel to determine the dose rates to various parts of the body. In a given situation, dose rates are grouped appropriately and compared with the three monthly Operational Controls. The dose rate which is highest in comparison with the related monthly Operational Control is used for personnel dose control in that situation.[3]

For simplicity in comparison of dose rates, the three dose rate groupings are divided by the numerical value of the related monthly Controls. The highest value will result from the calculation involving the dose rate of interest. For example, in a sample situation, the following dose rates might occur:

General gamma field   20 mR/hr

General beta field   10 mrad/hr

Contact hand dose rate on
    fuel element   250 mR/hr (gamma) +
                                     100 mrad/hr (beta)

Grouping this information according to the requirements of the Operational Controls, the following are obtained:

(1) Dose rate to the Whole Body   20 mrem/hr

(2) Dose rate to the Skin of the Whole
     Body   30 mrem/hr

(3) Dose rate to the Extremities
     (hands)   350 mrem/hr

To determine the relative importance of each dose rate for control purposes, they are divided by the numerical value of the appropriate Operational Control as follows:

(1) Dose rate to the Whole Body $\quad \dfrac{20 \text{ mrem/hr}}{1} = 20$

(2) Dose rate to the Skin of the $\quad \dfrac{30 \text{ mrem/hr}}{3} = 10$
Whole Body

(3) Dose rate to the Extremities $\quad \dfrac{350 \text{ mrem/hr}}{8} = 43.8$

In this situation, dose to the hands is relatively the most important, and would be utilized for personnel dose control purposes. Controlling personnel dose by routinely utilizing the most limiting dose rate assures that the dose to all parts of the body including the hands will be controlled within established limits.

DOSIMETRY

Hand dosimetry at Pacific Northwest Laboratory utilizes a lithium fluoride-in-teflon thermoluminescence element enclosed in a rubber finger ring.[4] This type of dosimeter has many advantages over the film dosimeter used previously. In particular, it has improved energy response and lower sensitivity to environmental factors other than radiation. Results are more consistent and can be correlated more readily with field measurements made with dose rate survey instruments. Data from these dosimeters are utilized for monthly control purposes as well as for personnel dose record information. However, it is realized that such dosimeters may not always record the dose at the point of greatest interest. They are normally worn on the finger near the palm of the hand. When personnel handle relatively large radiation sources, dosimeter results are indicative of dose to the entire hand, but when a very small source is handled, the tips of the fingers may receive appreciably more radiation then the dosimeter indicates. Special studies are often required to determine the dose rate and dose of interest in such cases.

WORK EXAMPLES

The usual situation resulting in significant extremity dose involves radioactive material sufficiently large and/or massive that it must be manipulated with the hands. In such cases, a finger ring dosimeter located near the palm will provide a representative measurement of the extremity dose. In some situations, physically small pieces of radioactive material are manipulated with tweezers and other such devices which ensure maintenance of some distance between the hands and surface of the material. In these circumstances, the additional distance from fingertip to finger ring is normally small by comparison with the distance between the material and the fingertips. Here, the finger ring dosimeter located near the palm also provides a representative measurement of the extremity dose.

Work involving exposure of the hands to large radiation sources was encountered at Pacific Northwest Laboratory during the program for development and fabrication of fuel elements for the Phoenix Program. The program involved fabrication of multiple-plate fuel elements containing Plutonium-239 as an $AlPu(20^W/o)$ alloy. The initial step in fabrication required melting and casting the alloy in the form of ingots. The ingots were then extruded into bars and the bars were cut into small billets. The fuel billets were inserted into individual cases of aluminum which were welded closed and outgassed. Hot rolling was then performed until plates of the proper thickness were obtained. Subsequent operations involved frequent quality control checks, especially during trimming and forming the fuel plates to their final shape

(Fig. 1). After fabrication of the individual plates they were cleaned, inspected and assembled into the Phoenix element (Fig. 2). Contact hand dose rates to a maximum of approximately 1.5 rem/hr were encountered during the program. Tools were used for manipulation whenever practicable, but when the use of tools was not possible the materials were manipulated by hand. The nature of the work was such that finger ring results were acceptably representative of the extremity dose, so that no special evaluation methods were required.

Occasionally, circumstances combine in such a way that the finger ring may not be representative of extremity dose if worn in the usual manner near the palm. This situation occurred recently in a research program at Pacific Northwest Laboratory involving x-ray diffraction analysis of various radioactive materials. In this program, the materials were loaded into glass capillary tubes measuring about 1/4 inch in length by 0.3 mm in diameter, with a wall thickness of 0.001 mm. Each capillary was then loaded into a second capillary tube whose internal diameter tapered from 0.5 mm to 0.3 mm with a wall thickness of 0.001 mm. These capillaries are extremely fragile, and the technicians who sealed the tubes and mounted them in the x-ray diffraction cameras initially felt that touch sensitivity was insufficient unless the capillaries were manipulated with the fingertips. This technique resulted in high fingertip dose which was not reflected in finger ring dosimeter results (Fig. 3).

Among the materials which have been analyzed in this program is promethium-147 oxide. In the analysis of this material, the capillary is loaded with 50 to 100 μgm of the oxide. Because of the high specific

Figure 1.
Handling Pu alloy fuel plates

Figure 2.
Assembling the Phoenix element

activity of promethium-147 and the beta emitted in its
decay, high fingertip dose was anticipated even with
short handling times. When the responsible health
physicist was first approached about this program,
his initial step was calculation of the probable fingertip dose. In estimating the fingertip dose, calculations included the effects of shielding afforded the
dermal layer of the skin by the epidermis and the
gloves which were to be worn. These calculations
indicated that a single thickness of surgeon's gloves
and the epidermis would reduce the dose to the derma
of the fingertips to relatively low levels. Procedures
were then established which would permit handling of
the capillaries with the fingertips as long as handling
time was kept to a minimum and at least one pair of
surgeon's gloves was worn. As soon as a capillary
loaded with promethium oxide was available, measurements were performed to check the accuracy of the
computed dose estimates. These measurements indicated that one thickness of surgeon's gloves and the
epidermis would reduce the dose rate to the derma to
no more than 0.0001 of the contact dose rate. For a
capillary containing 100 μg of promethium oxide, the
dose to the fingertips during a one-minute handling
period was determined to be about 25 mrem.

Although finger rings cannot directly provide an
accurate measurement of fingertip dose in situations
of this kind, their continued use was recommended
after review of the work. Factors leading to this
recommendation included the small fingertip dose per
capillary, the small number of capillaries handled
each month, the handling interference introduced by
fingertip dosimeters, and the increased possibility of

capillary breakage due to that interference, with probable release of promethium oxide to the work area. Corrections developed in this study were utilized to relate finger ring dosimeter results to fingertip dose.

While special evaluation of this work indicated that fingertip dose was not currently a problem, the inability to obtain accurate direct dose measurement was of concern to the involved technicians and the assigned Radiation Monitoring personnel. As a result, the technicians developed a new handling technique which virtually eliminated the measurement problem. They found that ordinary laboratory tweezers could be modified for this work by installing sponge rubber pads on the tips. This modification permits adequate touch sensitivity without fear of capillary breakage (Fig. 4). When the technician utilizes the modified tweezers, finger ring dosimeters provide an acceptable measurement of dose to the entire hand, including the fingertips. Following development of the modified technique, the applicable Radiation Work Procedures were amended to require handling loaded capillaries with tweezers. Due to this development, forseeable increases in workload are not expected to require extensive reevaluation of hand dose for this operation.

SUMMARY

The history of radiation work experience shows that the hands are the most common site of radiation injury. Radiological situations involving significantly higher dose rates to the hands than to the whole body

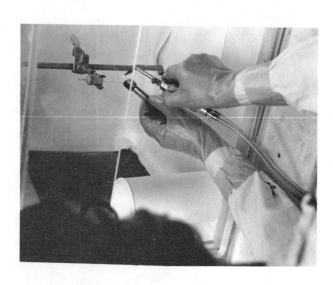

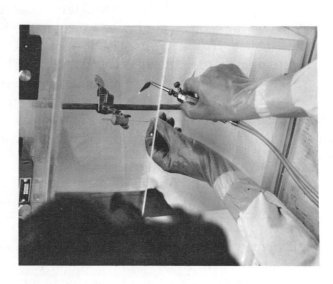

Figure 3.

Handling capillary tubes containing 147Pm produces high fingertip doses which are not adequately monitored

Figure 4.

The use of specially modified tweezers greatly reduces fingertip doses.

should be carefully evaluated. A technique is in use at Pacific Northwest Laboratory for evaluating the relative significance of dose rates to various parts of the body in any given work situation. The dose rate which has the greatest significance in a given work situation is utilized for personnel dose control in that case. This technique assures that dose to all parts of the body, including the hands, is maintained within established limits.

REFERENCES

1. B. Lindell, "Occupational Hazards in X-ray Analytical Work," Health Physics, 15, No. 6, pp. 481-486, December, 1968.

2. "Radiation Protection Procedures," BNWL-MA-6 (Procedures manual).

3. "Procedures for Radiation Monitoring," BNWL-MA-7 (Procedures manual).

4. L. F. Kocher, R. L. Kathren, and G. W. R. Endres, "Thermoluminescence Personnel Dosimeters at Hanford, I. ^{7}LiF Extremity and Non-Radiation Worker Dosimeters," October, 1968.

CONTROL OF PERSONNEL EXPOSURES TO EXTERNAL RADIATIONS IN A PLUTONIUM CHEMICAL PLANT*

J. B. Owen
The Dow Chemical Company
Rocky Flats Division
Golden, Colorado

ABSTRACT

This paper describes a practical approach to the control of personnel exposures to external radiations in a chemical plant handling kilogram quantities of plutonium in glove boxes. Calculated dose rates from various processing operations are compared to measured dose rates. Shielding materials for x-rays, gamma rays, and fast neutrons and their economic and operational impact are discussed. Personnel training, rotation, and work habits which must be considered in maintaining exposures below the accepted guide levels are included.

*Based on work performed under U.S. Atomic Energy Commission Contract No. AT(29-1)-1106.

INTRODUCTION

The intent of this paper is to describe the application of a practical approach to the control of exposure to external radiations during the chemical processing of kilogram quantities of plutonium in glove boxes. I will, therefore, limit the discussion of technical considerations to that which is necessary for the sake of understanding.

When one considers the decay properties of the plutonium isotopes 238, 239, 240, 241, and 242,[1] it will be noted that all are alpha emitters except 241 which is primarily a beta emitter. These properties, numerous animal experiments, and limited observation of human internal exposure have resulted in the acceptance of glove boxes as a means of limiting the internal exposure potential while handling plutonium.

Further study of the decay properties[2] will reveal that when these isotopes of plutonium decay, a small part of their energy is given off as x and gamma rays. Decay products, such as americium 241 and uranium 237, contribute significant amounts of additional x and gamma rays. Spontaneous fission results in neutrons and additional gamma rays and when the plutonium is mixed or compounded with light elements, additional neutrons are produced by alpha-neutron reactions.

The amount of x-rays, gamma rays, and neutrons from a mixture of plutonium isotopes is dependent upon the isotopic concentration and time since chemical purification. At Rocky Flats the majority of the plutonium mixtures handled have an isotopic concentration of less than 0.1 percent 238, about 93 percent 239, about 6 percent 240, less than 1.0 percent 241 and less than 0.1 percent 242. (These are mass, not activity, fractions.) The time since chemical purification ranges from a few days to more than ten years.

The discussion of personnel dosimetry techniques and radiation detection equipment used at Rocky Flats would constitute complete papers in themselves. I will merely mention that we are including thermoluminescence in our planning for the future, but we are presently using DuPont 558 beta-gamma film worn on the body and at the wrist and Kodak Type A neutron monitoring film worn on the body. All radiation surveys for x and gamma rays are performed with an ionization chamber, such as the Victoreen 440 or 440 RF, and all neutron surveys are performed with a Bonner Sphere type[3] survey instrument.

The Rocky Flats plutonium chemistry operations, which are performed in glove boxes and to which this discussion is directed, are batch dissolution of plutonium metal, oxide and tetrafluoride, continuous peroxide precipitation, americium separation and purification, continuous calcining, continuous fluorination, and batch reduction to metal. These operations involve hundreds of kilograms of plutonium and up to 100 grams of americium each month.

X AND GAMMA RAYS

H. A. Moulthrop[4] has reported the results of a study of the energy spectrums of the x and gamma rays from special plutonium samples in various physical forms and the effectiveness of various shielding materials, such as lead, lead glass, steel, safety glass, and Plexiglas.

Rocky Flats experience indicates that significant exposure to x and gamma rays can result from the handling of a few hundred grams of plutonium. We have found that the dose rates at the work stations can be reduced to less than 1.0 mR/hr through the use of 1/8 inch thick lead sheets over the metal portions of the glove boxes, 1/4 inch thick lead glass (equivalent to 1/16 inch of lead) over the windows, and hinged covers of 1/8 inch thick lead over the glove and bag ports. Hand exposures are controlled through the use of leaded dry box gloves. The majority of these gloves contain the equivalent of 0.16 mm of lead with a maximum of 0.36 mm of lead used at the operations which concentrate the americium 241.

Our experience indicates that this shielding has very little, if any, affect on the operator's efficiency. The cost of the glove box is increased by a factor of about 1.8 when this shielding is added.

NEUTRONS

When one considers neutron shielding, the energy of the neutrons must first be determined. The neutrons from spontaneous fission have a continuous energy spectrum with a maximum intensity near 1 MeV. The neutron energy and abundance from alpha-neutron reactions is dependent upon the alpha energy and the target nuclei. Rocky Flats experience indicates that for the types of operations under discussion the predominant sources of neutrons are spontaneous fission and alpha-neutron reactions in oxygen and fluorine. For practical purposes, all of these neutrons can be considered as having 1 MeV of energy.

F. J. Allen and A. T. Futterer[5] have reported the dose transmission factors for various neutron energies in polyethylene, water, concrete, and Navada Test Site soil. This data and studies at Rocky Flats indicate that for the neutrons of concern polyethylene has a half value of 1.2 inches and a tenth value of 3.3 inches, water has a half value of 1.25 inches and a tenth value of 3.5 inches, Plexiglas* or Benelex** have a half value of 1.8 inches and a tenth value of 4.9 inches, and concrete has a half value of 2.1 inches and a tenth value of 5.7 inches.

Calculations will show that the neutron dose rate will exceed 1.0 mrem/hr at the work station if the amount of plutonium exceeds about 4 kilograms as metal, about 2 kilograms as oxide, or about 20 grams as a fluoride.

Time will not permit a detailed discussion of the chemistry operations, but it should be noted that the health physicist must be aware of the detailed design of the processing equipment and watch for potential problems. For example, the design of the Rocky Flats continuous calciner involved an almost horizontal tube about 8 feet long with a loading hopper on one end and a receiving hopper on the other. It was intended that each hopper would contain up to 2 kilograms of plutonium and that the tube would contain up to 5 kilograms of plutonium all as oxide. A line source calculation indicated a neutron dose rate of about 1.8

---

*Manufactured by Rohm and Hass, Philadelphia, Pennsylvania, and American Cyanamide, Wallingford, Connecticut.

**Manufactured by Masonite Corporation, Chicago, Illinois.

mrem/hr would result at a point 2 feet from the center of the tube. Neutron surveys at this point indicated up to 6 mrem/hr. Investigation of this difference revealed that plutonium oxide had spilled from
the hoppers and a significant number of kilograms of
plutonium had collected in the drive mechanism which
turns the tube.

As a word of caution, criticality experts should
be consulted and made aware of any plans to install
neutron shielding. These shielding materials are
also very effective neutron reflectors.

At Rocky Flats we have found that the practical
limit for the thickness of neutron shielding on a glove
box is a maximum of 5 inches. The first 2 inches do
not have a significant effect on the operator's reach
and vision, but at 5 inches the operator is seriously
restricted. The cost of the glove box is increased by
a factor of about 2.2 when 2 inches of shielding is
added and a factor of about 2.8 by the addition of 5
inches of shielding.

PERSONNEL TRAINING AND ROTATION

The experienced health physicist will be quick to
point out that shielding alone is not a complete control.
The shielding merely makes it possible for the trained
operator to use the shielding in a manner which maintains exposures below the accepted guide levels. At
Rocky Flats we have found it necessary to maintain a
continuous training program. The training program
stresses the need for good housekeeping practices inside of the glove boxes, good storage practices

outside of the glove boxes, and the use of shielding, time and distance to limit exposures.

Supervision has found it necessary to administer a rotation program such that the workers are frequently moved between operations of high and low exposure potential. This program requires continuous and conscientious attention of the workers and their supervision.

CONCLUSION

The Rocky Flats experience indicates that kilogram quantities of plutonium mixtures typical to Rocky Flats can be chemically processed in glove boxes while maintaining external exposure within the accepted guide levels providing the proper shielding, personnel training and personnel rotation programs are utilized.

REFERENCES

1. C. M. Lederer, J. M. Hollander, and I. Perlmon, Table of Isotopes (John Wiley and Sons, New York, 1968).

2. W. C. Roesch, Surface Dose from Plutonium (USAEC Report HW-51317, 1957).

3. D. E. Hankins, A Neutron Monitoring Instrument Having a Response Approximately Proportional to the Dose Rate from Thermal to 7.0 Mev (USAEC Report LA-2717, 1962).

4. H. A. Moulthrop, Z. Plant Radiation Study In-
   terim Report No. 5—Part II, Data on Gamma
   Shielding of Special Plutonium Samples (USAEC
   Report HW-61755 PT2, 1959).

5. F. J. Allen and A. T. Futterer, Neutron Transmission Data (NUCLEONICS, Vol. 21, No. 8,
   August, 1963).

PLUTONIUM: PERSONNEL EXPOSURE CONTROL WITH INCREASING 240PU CONTENT*

G. D. Carpenter
Battelle Memorial Institute
Pacific Northwest Laboratory
Richland, Washington

ABSTRACT

In recent years there has been a marked increase in the use of commercially recycled plutonium in fuel element fabrication programs. Commercially recycled plutonium has a higher plutonium-240 content than does the plutonium which has been available in the past. This higher plutonium-240 content results in increased photon and neutron dose rates which are sufficient to require modification of exposure control practices and procedures. This paper discusses some of the modified procedures which have been put to use

---

*This paper is based on work performed under United States Atomic Energy Commission Contract AT(45-1)-1830.

at the Pacific Northwest Laboratory to effectively maintain personnel dose within the regulatory limits.

INTRODUCTION

For the past several years the plutonium handled across the nation has been composed largely of ^{239}Pu. During this time the main health physics problem associated with plutonium handling has been containment to ensure that plutonium was not released into areas where deposition within the body was possible. Today there is a rapid increase in the use of commercially recycled plutonium containing significant quantities of ^{236}Pu, ^{238}Pu, ^{240}Pu, ^{241}Pu, and ^{242}Pu. The presence of these radionuclides in no way lessens the importance of containment; rather, it may significantly increase both the importance and difficulty of maintaining containment. Added to the increased containment problem, however, is a new one—high photon and neutron dose rates. The photons are the X-rays associated with the plutonium and americium; the neutrons are those from the spontaneous fission of plutonium and from the α, n reaction with light nuclei. We have developed a number of procedures designed to cope with the problems of high dose rates associated with handling of commercially recycled plutonium.

PHOENIX

Commercially recycled plutonium has been handled at Pacific Northwest Laboratory in a variety of physical and chemical forms for several years in association with various research and development programs. The quantities of material have been small, generally less than a few kilograms. During these programs, due to the relatively short handling periods involved, the increased dose rates were such that only minimal exposure controls were instituted. More recently, however, Pacific Northwest Laboratory undertook the Phoenix Fuel Fabrication Program. The Phoenix

Program required development and fabrication of plu-
tonium-aluminum alloy fuel elements for use in the
Materials Test Reactor in Idaho.   The alloy contains
approximately 21 ^{W}/o plutonium.   The plutonium consists of about 0.5% ^{238}Pu, 65% ^{239}Pu, 23% ^{240}Pu,
7.5% ^{241}Pu, 3% ^{242}Pu, and a trace of ^{236}Pu.   Also
of interest from a radiation standpoint is that approximately 0.5% ^{241}Am has grown in from the beta decay
of ^{241}Pu in the two years since the plutonium was
purified.   Fabrication process development efforts
began about two years ago, and for the past year
element fabrication has been underway.

FABRICATION PROCESS

In the fabrication program the plutonium is first
alloyed with aluminum and cast into extrusion billets.
These billets are then extruded into a long bar which
is subsequently cut to form small squares.   The
squares are placed into an aluminum 'picture frame'
case which is welded together and outgassed.   After
outgassing, the encased alloy is hot rolled into plates
approximately 40 mils thick.   The alloy core is then
roughly located with radiation detection instruments,
and the plate is sheared to a smaller size.   The core
is precisely located by x-raying, and the plate is
trimmed to final dimensions.   The plates are then
stamped to the desired shape and assembled into elements.   Throughout all of these processes various
quality control samples are taken and the fuel plates
are inspected many times.   The process is considerably more detailed and complicated than indicated in
this brief description.

During development of the fabrication process,
little difficulty was experienced in controlling personnel dose.   Fuel quantities were initially small and

increased gradually as the process approached that
finally adopted for fabrication. As fuel quantities
and laboratory occupancy times increased, radiation
doses also increased. During this period of increasing radiation intensities, exposure control was maintained by use of 1/4 and 3/8 inch thick, leaded lucite
panels on the windows of the glove boxes and fume
hoods being used. With finalization of the process,
fabrication began in earnest. There was a marked increase in personnel doses at this time due to the
greatly increased fuel inventory and laboratory occupancy times. To minimize the doses to personnel,
several steps were taken. Leaded lucite panels were
now placed on all glove boxes and hoods being used in
the fabrication process. Lead-impregnated vinyl
sheets were used to cover all fuel when it was not
being actively worked. Sheet lead was formed into
covers for use where the lead-impregnated vinyl
sheets were not practical (Fig. 1). Special shielded
enclosures were assembled for use whenever the process step precluded the use of other shielded equipment (Figs. 2 and 3). Personnel were required to
wear lead-impregnated aprons when working with fuel
materials (Figs. 1, 2, and 3). Operating procedures
also required that fuel inventories within the fabrication laboratory be minimized to the extent practical.

ASSEMBLY INTO ELEMENTS

Through June, 1968 the neutron dose amounted to
about six percent of the gamma dose. To this point
the fabrication effort had been directed toward accumulation of a sufficient number of fuel plates to permit
assembly of all required elements. By July enough
fuel plates had been accumulated, and identification,
sorting, and cleaning of the plates began. To accomplish the identification and sorting, large numbers of

Figure 1.
Sheet lead shielding of a Phoenix element

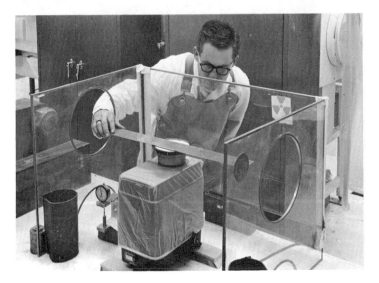

Figure 2.
Special shielded enclosures and lead
aprons are used to reduce doses.

Figure 3.
Leaded lucite panel shielding
provide important reductions in
radiation fields.

fuel plates were brought into the fabrication laboratory. The accumulation of fuel brought about a slight increase in the background photon dose rates in the laboratory.   The whole body neutron dose increased to about 15% of the photon dose during July, August, and September.   The increased neutron to gamma ratio occurred because of the increased quantities of fuel handled at one time.   For a given fuel form and configuration there is a geometry at which further increases in quantity do not materially increase the photon dose rates.   (The photons associated with the plutonium isotopes having low energies, self-shielding prevents additional increases in photon dose rates.)   However, the neutron dose rates increase roughly in proportion to the quantity of plutonium in the fuel present.   Since the identification, sorting, and cleaning operations were largely performed in the open and it was generally impossible to make use of the shields previously utilized, the whole body photon dose increased over that experienced earlier.

In October assembly of the elements began. There was again an increase in the rate of accumulation of both photon and neutron dose to personnel.   Neutron doses amounted to 50 to 100% of the photon doses.   This increase was due to the large number of fuel plates handled at one time and the increased photon and neutron background intensities.   As a result of the rather large and sudden increase in personnel doses, work was briefly suspended while the process was re-examined. The major contributions to the large increase in personnel dose were found to be the large fuel inventory in the laboratory work area and inadequate use of leaded lucite shields for close contact work Controls based upon these findings resulted in

large reductions in the whole body dose accumulation
rate and in the neutron to photon dose ratio. The
neutron dose dropped to less than ten percent of the
gamma dose where it remained throughout the rest
of the program. Whole body doses ranged from two
to five rems for the year for the personnel involved.

RECYCLED PLUTONIUM

Large quantities of commercially recycled plutonium are also handled in another program recently
undertaken by Pacific Northwest Laboratory. In this
program plutonium oxide powder, approximately 23%
of the plutonium being ^{240}Pu, is pressed into small
blocks. Photon and neutron dose rates from a given
quantity of commercially recycled plutonium as plutonium oxide are not as high as those from a similar
quantity of plutonium as the Pu-Al alloy utilized in the
Phoenix Program. The ^{240}Pu content is comparable
in both cases, but the photon dose rates at normal
working distances from the oxide are about 25% of
those from the alloy. Neutron dose rates from the
oxide are about 60% of those from the alloy. Nonetheless, dose rates are sufficiently high to warrant use
of the same exposure control methods and practices
as were used with the alloy.

RECOMMENDATIONS

As a result of the experience gained working with
commercially recycled plutonium in the described
programs, several recommendations for controlling
personnel exposure can be stated. Implementation
of the recommendations in any program involving
manipulation of commercially recycled plutonium
should enable reasonably easy and inexpensive maintenance of personnel doses within applicable regulatory
standards.

1. Minimize plutonium quantities in occupied areas.

> Minimization of fuel quantities will minimize photon and neutron dose rates. Dose rates and neutron to gamma ratios vary with fuel composition, geometry, and quantity, but quantity reduction is generally the fastest, easiest means of lowering dose rates. Fuel fabrication should therefore be aimed at maintaining the lowest reasonable quantity at each work station. Important also is minimization of the total inventory within the general laboratory but outside specifically designed storage enclusures. This minimizes background radiation levels within the laboratory and significantly reduces personnel dose.

2. Utilize photon shields.

> The experience gained at Pacific Northwest Laboratory indicates that reduction of the photon dose will do much to control personnel dose. Photon shielding should be installed on all glove boxes and hoods used in the program. Once the material is contained and cleaned and can be brought into the open laboratory area, shielding for all possible process steps should be constructed and used. Where enclosures or panels cannot be used, covering with sheet lead or lead-impregnated vinyl sheets is effective if carefully controlled. Personnel working with commercially recycled plutonium should be required to wear glasses and leaded aprons whenever working with the material.

3.   Constantly reexamine the program

   The program should be constantly reex-
amined to ensure that everything possible has
been done to minimize dose. When it is
asserted that the inventory of fuel in each process step and laboratory is the minimum which
must be maintained to perform the program,
examine that assertion closely. In the instance
of the Phoenix Program it was found that the
fuel inventory could reasonably and effectively
be reduced by proper planning and management. Reevaluate process steps which apparently cannot be shielded. Usually, full or
at least partial shielding can, in fact, be
accomplished, or it may be that these process
steps can be done by some method which will
increase the distance between worker and fuel.
Do not overlook the seemingly inconsequen-
tial parts of the process, such as movement
of fuel between laboratories, movement of
fuel between process work stations, or increased background dose rates in rooms adjacent to storage or process areas. Finally,
and perhaps most important, routinely ob-
serve operating personnel at work. Famili-
arity with the process often results in personnel
adopting practices which bypass the procedures
and equipment designed to minimize exposures.

4.   Evaluate neutron dosimeter calibration.

   Pacific Northwest Laboratory uses NTA
film for neutron dosimetry. Calibration is
performed with a PuF_4 source which has

proven satisfactory for several applications. However, it was determined that the PuF_4 calibration overestimated the dose for the neutron spectrum associated with the Pu-Al alloy used in the Phoenix Program.[1] Dosimeter accuracy for a variety of plutonium compounds and isotopic compositions is presently under study. According to these findings, the calibration of neutron dosimeters should be evaluated for each program involving work with commercially recycled plutonium.

It is clear from the experience at Pacific Northwest Laboratory that commercially recycled plutonium can be handled without exceeding regulatory dose limits for fuel fabrication processes such as the Phoenix Program. Maintenance of personnel doses within the limits can be accomplished without resort to hot cell type facilities but with the use of simple shields and minimization of plutonium inventories. Neutron shielding, while not necessary under the conditions of this fuel fabrication effort, warrants consideration.

REFERENCE

1. L. G. Faust, private communication, to be published.

SOME OBSERVATIONS
OBTAINED FROM A NATION-WIDE
IN VIVO COUNTING SERVICE

Geo. Lewis Hegelson
and
David E. Pollard
Helgeson Nuclear Services, Inc.
Pleasanton, California

ABSTRACT

Since most radionuclides encountered in the
nuclear industry are insoluble, in vivo counting provides one of the best methods for determining internal burdens of many radionuclides. This paper discusses findings from approximately 7000 counts during the first three years of a nation-wide mobile in-
vivo counting service. Co-60, the most frequently
observed contaminant, will probably contribute the
most lung exposure. Its effective half-life is generally
about one to two years. Zn-65 is present in people
working at reactors where admiralty metal is used.
It tends to "mask" other nuclides such as Co-60. It
causes other operational problems such as high dose

rates from waste containers. Using an 8" x 0.5" NaI (Tl) detector in a Shadow Shield Whole Body Counter the authors have been successful in routinely measuring as low as 20% of the maximum permissible lung burden (MPLB) of U-235 in a 40 minute count. If the true lung burden is 0.245 mg of U-235, one MPLB at 93% enrichment, the probability is 90% that in any one measurement the observed activity will be 0.245 ± 0.088 mg. Most industries have an excellent safety record, however, a few have significant internal dose problems. The in vivo records have proven very useful in discovering trends in plant contamination status, boosting sagging employee morale, and in proving to regulatory agencies that the health physicist is using the latest techniques for estimating internal dose. The legal significance of an in vivo counting record appears far superior to urinanalysis records because of the relative simplicity of interpretation.

I. INTRODUCTION

In 1960 Dr. Claude Sill[1] published a paper on the experiences at the National Reactor Testing Station in the field of internal dosimetry. He stated that "In three years of whole body counting of human subjects at the NRTS, over 95 percent of all the nuclides encountered could not be detected in 1,500 ml samples of urine, the mode of elimination being almost exclusively by way of the GI tract." Many health physicists have been finding this same fact to hold true for their facilities. For example, while working at a major industrial laboratory the writers encountered a number of circumstances in which it was quite evident that urinanalysis was a poor technique for estimating internal body burdens. In one particular instance, a person had accidentially inhaled about 0.7 microcuries of Co-60. All urine and feces were analyzed for the first two weeks. Whole body counts were made daily with a Shadow Shield Whole Body Counter.[2] In the first three days approximately 0.45 microcuries were excreted in feces but a one hour count of a 3 day composite urine sample placed directly on a 3 by 3 inch NaI(Tl) detector gave no indication of Co-60. From the whole body counts, however, it was possible to follow the retention of Co-60 and to calculate the subject's lung dose. The effective half-life was found to be approximately one year.

Because of the inadequacies of urinanalysis and the apparent advantages of whole body counting, a company was founded to provide a mobile whole body counting service to the nuclear industry. This paper presents a summary of whole body counting results during the first three years of operation.

II. DESCRIPTION OF EQUIPMENT

Whole body counting was performed with the Shadow Shield Whole Body Counter which is manufactured by Helgeson Nuclear Services. For most fission and corrosion product work the detector is a Harshaw Chemical Company "Matched Window" Line Integrally Mounted 8" dia. by 4" thick NaI(Tl) Crystal-Photomultiplier with a resolution of 7.6% for Cs-137. For low energy work such as U-235 the detector is a Harshaw Chemical Company "Matched Window" Line

Integrally Mounted 8" dia. by 1/2" thick NaI(Tl)
Crystal-Photomultiplier with a resolution of 8.1% for
Cs-137. The detector signals were fed to a Nuclear
Data Model 130-AT 512 Channel Pulse Height Analyzer
which is normally calibrated for a nominal 10 keV/
channel. Readout is accomplished digitally with a
Tally Model 420 PR Tape Perforator, and IBM Computer Typewriter, and a Model ASR 33 Teletype;
graphically with a Varian Model F80 X-Y Recorder,
and visually with a Fairchild Model 701 Oscilloscope.
Data were read into a Digital Equipment Corp. Model
PDP-8/S 4096 word (12bit) Computer using a Tally
Model 424 PR Tape Reader. This equipment, except the computer, is contained in an air-conditioned,
insulated and carpeted 24-foot van. Figure 1 shows
an outside view of the trailer.

The Whole Body Counter uses a Bed Position Indicator, which allows one to plot counting rates from
the detector, providing qualitative localization of the
radioactivity. It is especially useful as a guide in
determining if an individual might have had contaminated clothing. The counter is also equipped with a continuously variable Automatic Analyzer Dead Time-
Bed Speed Correction Amplifier which keeps the bed
movement synchronized with the live time of the analyzer.

III. CALIBRATION

The detector and analyzer were calibrated for
channel versus energy using the technique of Heath
described in IDO-16880.[3] Energy calibration was
performed at least once daily using Cs-137 and Zn-65

Fig. 1. Exterior View of Mobile Whole Body Counter.

sources to assure that the energy-to-pulse height calibration matched that given by Heath. Typical results showed that the energy calibration was within an average of 0.5 channels over the energy range of 0.2 to 2.6 MeV.

The system was calibrated for sensitivity to various radionuclides by placing 200 small uniform activity sources in sugar phantoms ranging in weights from 27 pounds to 270 pounds. The sources were uniformly distributed in the phantom since it has been shown by previous work that this distribution provides results which adequately reflect the results obtained with non-uniform distribution such as might be found with insoluble particulates deposited in the lung. The calibration factors vary with body weight as do the Compton Scatter factors. In most cases these calibration factors may be represented as the sum of two exponential components. Figure 2 shows typical calibration curves.

IV. TYPES OF FACILITIES USING THE MOBILE WHOLE BODY COUNTING SERVICE

The acceptance of the mobile whole body counting service has been very good. The service has been provided to thirteen of the fifteen operating power reactors (although three of these have since been shut down), four of the seven major naval shipyards, three of the large radiochemical manufacturing companies, four of the major manufacturers of nuclear power reactors, a manufacturer of nuclear fuels, a nuclear fuel reprocessing facility, and many smaller organizations, universities, and several governmental

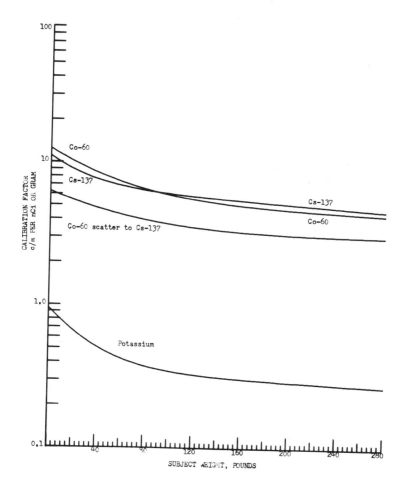

Fig. 2.   Typical Calibration and Scatter Factors.

installations. The trailer has been in all but six states
of the continental United States. Over 7,000 whole
body counts have been performed during this period
of time. The trailer makes from three to four trips
around the United States each year. A second unit
will be added in the spring of 1969. By 1972 there
will be at least four units scattered throughout the
country.

The frequency of counting is a function of the
probability of exposure of the personnel and of the
commitments which have been made in the license
applications. Generally speaking, the people at a
power reactor starting up for the first time will be
counted prior to actual loading of fuel and at yearly
intervals from then on until the contamination levels
start building up. After the reactor has been in operation for several years, and the contamination levels
are greater, the frequency changes to twice a year.
Some facilities are being counted on a quarterly basis.
Where the contamination levels are variable, a semi-
annual frequency appears to be the best compromise
between a desire to obtain a good estimate of the person's internal exposure and the economics of counting.
Most facilities count all of their people at least once
per year. This includes secretaries, laborers, administrative staff, etc., to assure that there has been no
inadvertant deposition. These people also serve as
"blanks" for quality control. If the counting frequency
is twice yearly (or more), the second count includes
only the radiation workers.

V. FINDINGS

Table I shows a list of the radionuclides which have been observed in three years of counting. Of these thirty radionuclides the only ones which have been reported as having been observed in urine are I-131, Hg-203, Se-75, Mo-99, Tc-99M, and Cs-134 and 137.

The most frequently encountered radionuclide is Co-60. This radionuclide has been found at every power reactor facility which has been in operation for any appreciable period of time, at all hot cell facilities, in all shipyards, at test reactors, a fuels reprocessing facility, and at several locations where repair work is done on pumps used in nuclear reactors. From the informal report of health physicists at locations where we have counted, no cobalt has ever been found in urine. (This same statement may be made about almost all the radionuclides which we have observed.) Table II shows the probability of observing Co-60 and the ranges observed at a number of nuclear facilities.

At power reactors where Admiralty metal is used in parts of the primary coolant system, zinc-65 is always found. Although the primary coolant system is allegedly leak tight, there will always be some small pin-hole leaks through gaskets, around valve packings, etc. The zinc, which is a component of Admiralty metal, is corroded by the high purity water, passes through the reactor core, and becomes activated. The zinc-65 rapidly becomes distributed throughout the controlled areas of the plant. Although the total body is the critical organ for zinc-65, it is not excreted in the urine. Therefore, whole body counting is the only

Table I
Nuclides Found in Past Three Years

Nuclide	Min. Sen. nCi	Nuclide	Min. Sen. nCi
Americium-241	0.5	Mercury-203	6
Antimony-124	2	Molybdenum-99	12
Antimony-125	7	Potassium	20 grams
Barium-Lanthanum-140	2	Protoactinium-233	4
Cerium-141, 144	27	Ruthenium-103, 106	13
Cesium-134	2	Selentium-75	6
Cesium-137	3	Silver-110m	2
Chromium-51	31	Strontium-85	3
Cobalt-58	2	Tantalum-182	2
Cobalt-60	2	Technetium-99M	2
Gold-198	3	Tin-113	3
Iodine-131	3	Tungsten-181	4
Iridium-192	2	Uranium-235	0.05 mg
Manganese-54	2	Zinc-65	4
Mercury-197, 197m	40	Zirconium-Niobium-95	3

effective means of measuring the internal dose from
zinc-65.　Figure 3 shows the distribution of body burdens found at a power reactor as a function of time.
For clarity, the individual data points have been represented as continuous lines.　Note that at the time of
the first count the body burdens ranged from a low of
6 nanocuries to a high of 2,000 nanocuries.　From
September 1965 through July 1966 the median body
burdens remained approximately the same at around
80 to 90 nanocuries.　In the fall of 1967 the Admiralty
metal was removed from the reactor system.　The

Table II

Probability of Observing Cobalt 60 at Power Reactors

Median Body Burdens, 95% "Ranges"

Reactor	Probability	Co-60, nCi Median	5% Above	5% Below
A. 9/65	25/75	16	110	7
12/65	38/70	8	22	3
7/66	5/44	23	45	11
1/67	4/64	4	6	2
5/67	20/62	1	6	0.6
11/67	3/63	20	40	10
5/68	5/62	6	6	6
12/68	5/60	16	21	12
B. 11/66	56/64	3	12	1
5/67	10/41	2	3	1
9/67	10/70	5	9	3
4/68	15/62	6	11	3
9/68	9/66	6	11	3
C. 12/66	0/51	-	-	-
9/67	33/53	7	15	3.5
9/68	20/37	15	50	5
D. 11/66	37/104	9	26	3
3/67	105/109	8	32	2
6/67	59/67	3.5	11	1
9/67	90/90	11	31	4
11/67	80/82	7	25	2.5
2/68	104/154	10	21	9
6/68	64/95	12	38	4
9/68	100/100*	11	18	6
11/68	43/112	12	25	6

*Highly variable background caused difficulty in making proper evaluations at low levels of Co-60.

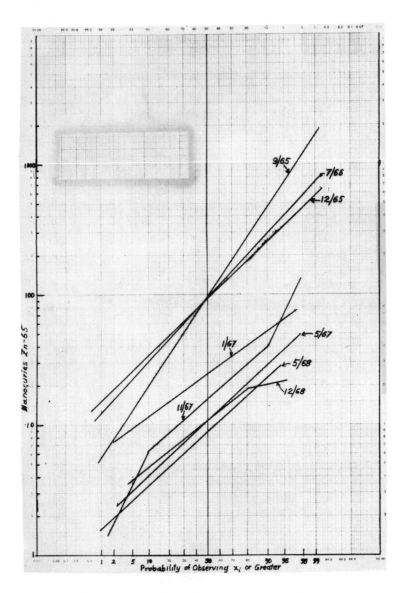

Fig. 3.   Zn-65 at Reactor "X".

resulting decrease in body burdens is markedly shown in the distributions. The most recent series of counts made in December 1968 show the zinc is only found in small quantities. Figure 4 shows how the total body burdens in units of "man-nanocuries" for all employees has changed versus time. This demonstrates that once the problem was recognized, action was taken to minimize it.

Even if the radionuclide is soluble in body fluids and a known fraction of it is excreted in the urine, evaluation of excretion data can be very complex. The health physicist at a radiochemical manufacturing facility wished to obtain a correlation between urinanalysis and whole body counting. Twenty individuals who were working routinely with various radiochemicals were selected for the program. Each person was requested to submit a urine sample at the beginning of the work day for a two week period. The people were whole body counted on the Monday of the second week. A typical graph of the I-131 excretion rates as a function of time for one individual is shown in Fig. 5. These wide variations were undoubtedly caused by the widely varying intakes which occurred during the study period. The whole body count showed approximately 600 nCi (out of a permissible body burden of 700 nCi). Using an effective half-life of 7.5 days the excretion rate would have been predicted to be 40 nCi/liter. The conclusion was reached that whole body counting presented the only reliable estimate of the body burdens.

As a group, radiochemical manufacturing facilities have shown the greatest frequency for internal contamination and have also shown the highest body burdens. This summarized in Table III for the radionuclides I-131 and Hg-203.

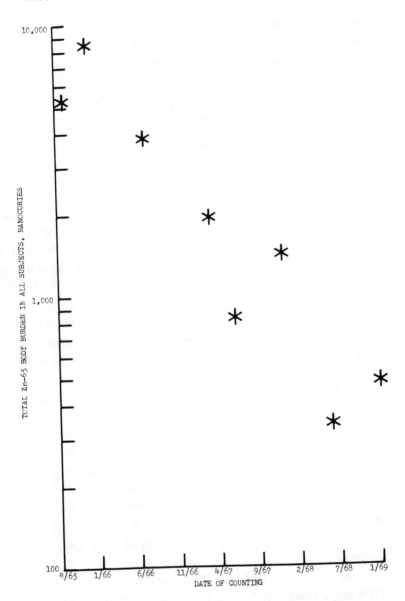

Fig. 4. Variation in Total Zn-65 Body Burden
vs Time at Nuclear Power Reactor "X".

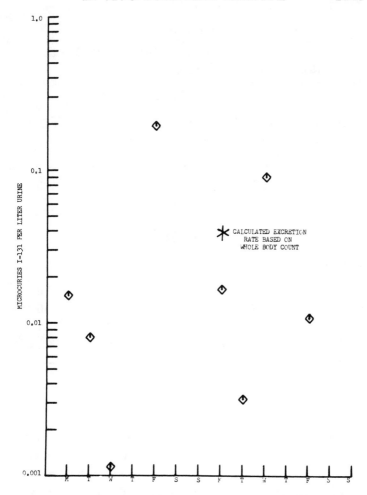

Fig. 5.   Bioassay Results vs Time, Radiochemical Manufacturing Facility.

Table III
Probability of Observing I-131 and Hg-203 at
Radiochemical Manufacturing Facilities

Iodine-131, nCi				
Company	Probability	Median	5% Above	5% Below
A	19/19	60	300	13
B	21/21	80	900	7
C	19/31	30	300	3
D	21/31	25	230	3

Mercury-203 nCi				
Company	Probability	Median	5% Above	5% Below
A	19/19	300	3000	30
B	21/21	200	800	50
C	21/31	50	450	7
D	15/31	70	180	22

VI. RECENT ADVANCES

With the addition of an 8" diameter by 1/2" thick crystal it has been possible to measure U-235 in the lung. Figure 6 shows a graph of an individual containing 320 micrograms of U-235. The data points are marked with the "x" while the computer calculated curve is a smooth line. Note that the smooth curve, calculated by a non-linear curve fitting program, matches the data points very well.

The limit of sensitivity based on a forty-minute count is approximately 50 micrograms of U-235. At one maximum permissible lung burden of fully enriched U-235 the over-all measurement error is

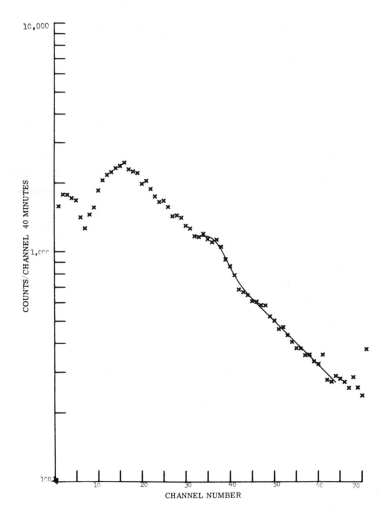

Fig. 6.   0.32 mg U-235 in Human Lung
X = Measured Value
———— = Calculated Curve.

approximately ± 40% at two standard deviations. This compares favorably with that reported by Cofield.[4] Calibration factors as a function of chest thickness are represented by a single exponential function.

In the manufacture of plutonium a small amount of plutonium-241 is almost always found along with the much larger amounts of plutonium-239. Plutonium-241 decays by beta emission to americium-241 which is a strong alpha emitter, but also emits a 60 kilovolt gamma which is measured with the 1/2" thick crystal. (The crystal was chosen for U-235 not Am-241. Ideally, a 2-4 mm crystal should be used for Am-241 measurements.) The minimum sensitivity is approximately 0.5 nanocurie of Am-241 with a two-sigma error of approximately ± 0.5 nanocuries. Analysis of the data is accomplished in a manner similar to that used for U-235. Because the area of integration includes the 50 keV iodine escape peak, the calibration factor is the sum of two exponentials as a function of chest thickness.

Much work remains to be done in improving sensitivity and reproduceability. However, an important beginning has been made and the technique is adequate for routine use at plutonium laboratories where the activity ratios of Pu-239 to Am-241 are relatively low. For example, Jech[5] reported a typical ratio of 13:1 at Hanford. Thus, with a minimum sensitivity of 0.5 nCi for Am-241, one could measure about 6.5 nCi equivalent of Pu-239, slightly less than one-half of a maximum permissible lung burden of 16 nCi.

VII. VALIDITY AS LEGAL RECORD

To the writer's knowledge the validity of an in vivo counting record has not been tested in a court of law, although in at least one case it constituted a very important part of the documentation of "no radio-activity due to employee's work" in a workmen's compensation trial. An in vivo count can reveal what radionuclides are present, and, due to the absense of photopeaks in certain channels, it can also reveal what radionuclides are not present above a given level. Table IV shows a typical table of minimum sensitivities. "Minimum Sensitivity" is defined as follows:

$$\text{Min. Sens.} = \frac{3\sqrt{(\text{Bkg c/m}) + (\text{K-40 scatter. c/m})}}{(\text{Counting time, min})}$$

$$+ \frac{\sqrt{(\text{Cs-137 scatter, c/m})}}{(\text{Calib. Factor})}$$

This type of documentation not only provides a good legal record but also plays an important part in employee morale. By looking at his spectrum in the oscilloscope or on the X-Y graph and by comparing what he sees with typical background and calibration error, the subject can see for himself that in most instances his radioactivity is essentially the same as a background count. Plant managers have frequently commented on the ready acceptance of in vivo counting by the employee and of the significant improvement in employee morale and trust in the health physics staff and program.

Table IV

Table of Minimum Sensitivities for Eight Minute Counts Based
on Background of 1/20/69 @ 0735 and 140 Grams of Potassium
Plus 10 NCI of Cesium 137 Number of Isotopes = 24
Length of Background Count = 25.55 Minutes

Radionuclide	Calib. Factors	Scatter Potass.	Factors CS 137	Back- Ground	Sensitivity	MPBB(1)	Units
Potassium.	0.45	0.000	0.00	118.35	25.6	None	Grams
Antimony 124	3.02	0.013	0.00	42.77	2.3	900(L)	NCI
Antimony 125	2.40	0.130	3.39	239.17	7.5	3200(L)	NCI
BA LA 140	4.85	0.219	0.00	66.81	2.2	600(L)	NCI
Cerium 141	0.98	0.700	4.11	648.84	30.4	650(L)	NCI
Cesium 134	6.22	0.122	0.96	200.78	2.6	20000(TE)	NCI
Cesium 137	5.73	0.120	0.00	213.97	2.8	30000(TB)	NCI
Chromium 51	0.75	0.272	3.02	511.23	34.0	600000(L)	NCI
Copalt 58	6.20	0.121	1.02	197.96	2.6	2900(L)	NCI
Cobalt 60	6.78	0.270	0.00	123.52	2.0	1100(L)	NCI
Gold 198	7.07	0.213	3.27	440.35	3.4	20000(K)	NCI
Iodine 131	8.35	0.311	3.85	382.97	2.7	700(THY)	NCI
Iridium 192	8.77	0.234	2.22	269.00	2.2	1400(L)	NCI
Manganese 54	6.67	0.130	0.13	206.77	2.4	3600(L)	NCI
Mercury 203	5.11	0.244	2.98	738.82	5.9	4000(K)	NCI
Ruthenium 106	1.57	0.192	2.45	315.02	12.9	600(L)	NCI
Selenium 75	3.35	0.110	1.10	324.26	5.9	8900(L)	NCI
Silver 110M	5.74	0.116	0.00	194.16	2.7	1000(L)	NCI
Strontium 85	6.72	0.152	2.00	315.02	3.0	5200(L)	NCI
Tantalum 182	5.14	0.144	0.00	199.80	3.1	1500(L)	NCI
Technetium 99M	14.42	0.737	5.66	627.04	2.1	200000(TB)	NCI
Tin 113	5.20	0.104	1.08	245.40	3.4	3600(L)	NCI
Zinc 65	2.63	0.140	0.00	171.35	5.6	60000(TB)	NCI
Zircon Niob 95	5.84	0.123	4.27	205.55	3.0	1600(TB)	NCI

(1) MPBB = Maximum Permissible Body Burden. The critical organ is denoted by the
letter(s) in parenthesis. L = lung, TB = total body, THY = thyroid, K = kidney.

VIII. SUMMARY

Whole body counting is probably the only reliable
and inexpensive method to measure most of the radionuclides encountered in the industries associated with
nuclear fuels and reactors due to the insolubility of
the dusts found in these facilities.

REFERENCES

1. Claude W. Sill, Jesse I. Anderson, and D. R. Percival, "Comparison of Excretion Analysis with Whole Body Counting for Assessment of Internal Radioactive Contaminants", published in "Assessment of Radioactivity in Man", pp. 217-219, I. A. E. A. , (1964).

2. H. E. Palmer and W. C. Roesch, "A Shadow Shield Whole Body Counter", Health Physics, 11, pp. 1213-1219, (1965).

3. R. L. Heath, "Scintillation Spectrometry Gamma-Ray Spectrum Catalogue", Vol. 1 and 2, Clearinghouse for Federal Scientific and Technical Information, (1964).

4. R. E. Cofield, "In Vivo Gamma Counting as a Measurement of Uranium in the Human Lung", Health Physics, Vol. 2, pp. 269-287, (1960).

5. John J. Jech, "Assessing the Probable Severity of Plutonium Inhalation Cases", presented at the 1968 Health Physics Society Meeting, Denver, Colorado, June 1968.

DEVELOPING PERSPECTIVE CONCERNING*
CONTAMINATION RELEASES

G. E. King and W. L. Fisher
Battelle Memorial Institute
Pacific Northwest Laboratory
Richland, Washington

ABSTRACT

The need and ability to detect and measure radioactive contamination are affected by several factors.
Location and magnitude of the contamination, particle
size, relative hazard of the isotopes involved, specific activity, and other aspects of the contaminant
should be considered, if possible, in evaluating the
seriousness of contamination releases. Instrumentation, monitoring techniques, monitoring costs, and
personnel relations all are affected by such considerations. Frequently, as much attention is paid to inno-
cuous situations as to those of some radiological

---

*This paper is based on work performed under
United States Atomic Energy Commission Contract
AT(45-1)-1830.

significance. This paper will generally discuss the
subject of contamination monitoring and will emphasize the need for placing each contamination release
in proper prespective.

INTRODUCTION

Of the various types of incidents that commonly
occur in our business, the release of contamination
is most likely to quicken the pulse of the operational
health physicist. Contamination releases might
range from very small to very large in terms of
radioactivity, from confined within a laboratory to
unconfined in the environs, and from harmless to
hazardous.

The operational health physicist frequently must
predict beforehand or determine after the fact the
nature, extent, and severity of contamination releases.
When a release occurs, he must have the faculty to
assess the situation quickly. Many people turn to him
for answers to such questions as: How extensive and
hazardous was the release? What caused it? Is it
under control? When may we get back to work? How
costly might it be? What is the public relations
significance? Is it reportable to the appropriate
regulatory agencies?

Given enough time, it is not difficult to obtain
answers to such questions. Frequently, however,
answers are due before sufficient time has elapsed
to allow the best answers to be determined. The
health physicist is then faced with the task of making
the greatest use of a minimum amount of information.

It is for such situations that the health physicist
must develop perspective in order not to overstate or
understate the significance of the release. Perspective is defined as the "capacity to view things in their

true relations or relative importance." It can be developed only if the health physicist has an appropriate set of criteria against which to compare his knowledge. Some criteria—maximum permissible exposures, for example—have been established by regulatory agencies or recommended by advisory groups. In some areas, however, criteria have not been established and the health physicist is left pretty much to his own devices. With or without the benefit of generally accepted criteria, the health physicist must develop perspective by preconceiving ideas of what might and might not constitute a problem under a variety of contamination release situations.

For example, consider the subject of surface contamination, complete with its many variables, such as: toxicity; specific activity; type and energy of radiation; particle size; nature and location of the contaminated surface; occupancy or use of the contaminated surface; air movement over the contaminated surface; and methods of measuring surface contamination. It is doubtful that this subject will ever be understood well enough to allow rapid, precise evaluation of surface contamination hazards. Yet, the health physicist must be prepared to evaluate this hazard in order to answer the questions raised by his management and others. He can accomplish this only if he has preconceived, reasonably justifiable ideas—that is, perspective—on the subject.

The following are some of our experiences and thoughts on the subject of developing perspective.

A MINOR RELEASE OF CERIUM-144

The need for perspective can be illustrated by a minor incident that occurred at a Pacific Northwest Laboratory facility in the fall of 1967. Approximately

ten millicuries of cerium-144 and promethium-147
attached to large particles of inert material was
released through a stack. The particles, dispersed
over an area of about 30 acres, were quickly shown
not to constitute a hazard to personnel. Being present in sufficient number and having sufficient radioactivity to be detected easily, however, they constituted a significant nuisance. That is, unless we
disposed of them we might not in the future be able
to distinguish between a fresh release and this old
cerium release. After trying unsuccessfully to wash
the particles into the soil, we finally resorted to
covering them over wherever possible and picking
up the remainder one-by-one. The cost of this operation was incurred not because of the hazard presented
by the cerium particles themselves, but because they
could have interfered with our ability to detect future
releases. Our perspective on the subject of particle
size was further broadened by the realization that the
amount of cerium released could not even have been
detected, using normal methods and instruments, had
it been contained in much smaller particles distributed
over the same area.

A MAJOR SPILL

At the other extreme, a recent laboratory spill
of liquid containing about 1000 curies of cerium-144
and other isotopes has provided ample opportunity to
broaden our perspective concerning large releases
within the facility involved. The measurement of dose
rates and personnel exposure, containment of the
release, and decontamination all presented problems
much different from the low level environmental release. In contrast to the nuisance caused by low level
particles in the environs, this large spill created
within the facility high dose rates and contamination

levels that had to be dealt with very cautiously. Caution and careful planning permitted us to recover safely and efficiently from this spill.

PREPAREDNESS FOR INCIDENTS

The health physicist's thoughts concerning contamination releases should begin long before a release occurs. Ideally, such thoughts should occur when facilities, equipment, and experiments are in the planning stage. Besides determining what could happen and how to prevent it, he should prepare himself and others to cope with releases should they occur. A few considerations essential in such preparation are discussed briefly below.

Of first consideration are the factors associated with the loss of control. Releases may result from any number of situations, such as a dropped source, a bag or glove failure, a filter failure, or a spill. Situations such as a dropped source or a spill are generally characterized by a sudden loss of material with no further prospect of additional release. Such releases require immediate confinement of the radioactivity to as small an area as possible. On the other hand, situations such as a filter failure may be characterized by a continued release of material until the source of the problem is located and controlled. Locating the source may be very time consuming, and once it is located, a course of action to stop the release must be selected. Meanwhile, a considerable area may be contaminated—as, for example, in the case of a release of radioactive material from a stack. Both the single release and the continued release may result in either severe or only nuisance level contamination. However, the extended release, even if a minor one, generally causes more concern, because no one knows what will happen next.

For example, the previously mentioned stack
release in the fall of 1967 seems to have occurred
during a period of several days before it was discovered. Although the release was quickly shown to
be minor, we were very concerned about the possibility of a more serious release occurring before the
source could be located and controlled. Fortunately,
no significant release of material occurred while the
source was being located. The source was determined
primarily from the composition of the material released and a knowledge of the locations where this
type of material was handled. In this instance, stack
samplers revealed no information, since the large
particles deposited on the walls of the sampling lines
before reaching the filter media.

Frequently the location of the release may be a
very important factor. An area seldom occupied and
easily sealed off presents a small problem compared
to an area of high occupancy with access not easily
restricted. It is clearly recognized that a release
of a small amount of radioactive material into a work
area where protective clothing is routinely worn and
the air is continuously monitored will probably not
result in a significant problem, while the same
amount of material released into an uncontrolled
hallway may result in a serious problem. One location is established for the eventuality of a loss of
control, while the other is not. Location considerations also apply to environmental releases. At
Pacific Northwest Laboratory, the laboratories are
located several miles from a population center, while
at other sites, potential release points may be located
much closer to populated areas. Consequently,
comparable releases at the two locations may result

in significantly different problems. For example, the
public relations aspects of a release are generally
more important for a site located close to populated
areas.

The amount of radioactivity present and the way
in which it is distributed also have a great bearing on
the severity of the hazard. If the material is distributed over a large area in the environs, the total radioactivity may be considerable but the hazard presented
may be very small. The same amount of radioactive
material released in a building, however, may present a serious hazard, and controlled access to the
building may be essential.

In addition to forming some thoughts on such
matters as the source, location, and size of possible
releases, one also must consider the nature of the
potential contaminants. The isotopes involved,
particle size, and chemical form are all subjects
worthy of some thought.

RELATIVE HAZARDS

Morgan[1] and others have investigated the relative
hazard of the important radionuclides. Such hazard
characterizations generally are applicable to day-to-
day work with radioactive material, including occasional, minor releases. They may not, however,
be directly applicable in the consideration of large releases. Further, they do not account for differences
in the difficulty of detecting and measuring similar
releases of various isotopes. Certainly the difficulty
of detecting and measuring a release is an item worthy
of consideration. For example, since plutonium-239
alpha particles are more difficult to detect than stron-
tium-90 beta particles, this knowledge could well be
an important factor in developing perspective concerning releases of the two isotopes.

Particle size also deserves consideration. While it may not be possible to predict the particle size distribution of a postulated release, thought can be given to the effects that large or small particles might cause. Respirability is the most obvious and probably the most important such effect. Naturally, size affects the mobility of the particles, and thus the extent of the release and the difficulty of containing it. Also, particle size and distribution have a significant and rather interesting effect on the ability and need to detect and measure contamination. While small particles are more hazardous from an internal deposition viewpoint, large particles might produce a greater external radiation hazard. Further, as we experienced in the minor cerium-144 release mentioned earlier, large particles can present a definite nuisance.

The chemical form of material released can be as difficult to predict as particle size. Frequently we can do little more than guess as to whether the material might be soluble or insoluble. Where better estimation of chemical form is possible, its effect on hazard, detection, and decontamination should be considered.

Certainly there are other factors to be considered in developing perspective in addition to those already mentioned. Actually contamination control for each location must be considered on a separate basis, for there are details associated with each operation, facility, location, and organization which tend to make the situation faced unique. Therefore, every operational health physicist should give serious consideration to all of the variables that affect the nature, extent, and severity of contamination releases within the facilities and environs for which he is responsible.

CONCLUSION

This paper has included no data, nor has it concerned new development in technique or instrumentation. For that matter, it has not even concerned new ideas. Rather, it has consisted of a brief discussion of simple and seemingly obvious matters. Its purpose has been to remind the operational health physicist to develop perspective concerning contamination releases—for if he cannot view them in perspective, who can?

REFERENCE

1. K. Z. Morgan, W. S. Snyder, and M. R. Ford, Health Physics 10, 151 (1964).

EXPOSURE AND CONTAMINATION CONTROL
IN A HIGH-LEVEL ^{147}PM LABORATORY*

Nathaniel A. Greenhouse and William J. Silver
Lawrence Radiation Laboratory
University of California
Livermore, California

ABSTRACT

A research program at the Lawrence Radiation Laboratory, Livermore, involved handling of kilo-curie quantities of promethium-147 in the development of radioisotope power sources. A high-level plutonium laboratory was reorganized for this work. The low energy of the ^{147}Pm beta necessitated development of special techniques for continuous air monitoring and contamination control. An LRL-designed air monitor and a shoe counter using large-area G. M. detectors are described. Assembly of the power sources involved delicate work which could not be

*Work performed under the auspices of the U. S. Atomic Energy Commission.

done remotely. Hand exposures from bremsstrahlung and gamma-emitting impurities became the limiting health physics problem. Experiences in exposure and contamination control are discussed in this paper.

INTRODUCTION

The availability of promethium-147 in relatively pure form from Pacific Northwest Laboratories has considerable research into its applications. Much of the research effort has been aimed at the development of radioisotope power sources. These devices typically contain multi-curie quantities of ^{147}Pm, usually as the relatively insoluble oxide.

The research program at the Lawrence Radiation Laboratory, Livermore, involved handling kilocurie quantities of ^{147}Pm$_2$O$_3$ which presented several problems relating to inhalation hazards, external dose limitation, and contamination control.

DESCRIPTION OF LABORATORY FACILITY

The material to be used was a relatively insoluble particulate, and it was decided to reorganize an existing high-level plutonium laboratory for this work. The building ventilation system operates on a once-through basis, with room air entering from a central corridor and exhausting through high-efficiency filters to the outside.

Fixtures were available for the installation of a gloved box line in which the power sources could be

assembled.   In addition, a hood with filtered exhaust
was available for decontamination of assembled sources
(Fig. 1).

Normal access to the laboratory was via the central corridor, although a crash door opening directly
to the outside was available for use in emergencies.

The plutonium solid waste recovery and the
promethium facilities shared the same laboratory.
Although plutonium waste recovery operations were
minimal during the promethium work, the waste recovery gloved box line was highly contaminated and
produced a background level of a few mR/hr in the
room.   This background radiation, mostly from ^{241}Am,
significantly affected the limits of sensitivity of the
promethium monitoring instruments.

A central vault in the building was used for storing radioactive materials while not in use in the
laboratory.

RADIOLOGIC CHARACTERISTICS OF ^{147}PM

Promethium-147, a pure beta emitter, emits
betas having an endpoint energy of 0.224 MeV and an
average energy of about 0.070 MeV.   These energies
correspond to ranges of about 50 and 7 mg/cm^2,
respectively.   The ^{147}Pm is produced from the beta
decay of the fission product ^{147}Nd (Fig. 2), at the
rate of about 3 curies per thermal megawatt-day
(based on aging of the reactor fuel for one half-life).
Promethium oxide has a specific power of 0.27 W/g.

The radiation associated with promethium is due
to the ^{147}Pm beta emission and bremsstrahlung, and
to gamma emission from the principal impurities,

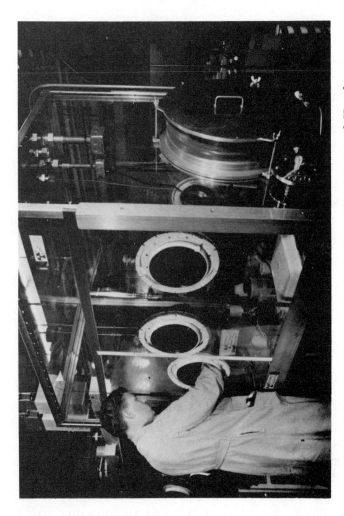

Fig. 1.   Promethium Gloved Box Line and Hood.

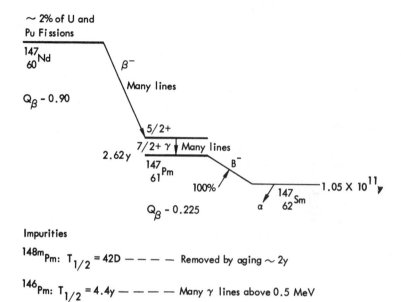

Fig. 2.   Production of ^{147}Pm.

^{146}Pm and ^{148m}Pm.   Light shielding will absorb the betas and most of the associated bremsstrahlung from ^{147}Pm, but the hard gamma emissions from the 146 and 148m isotopes require heavy shielding.   Fortunately, the 42-day half-life of ^{148m}Pm allows for its significant reduction by aging the material for about 2 years after separation.   Because of its longer half-life (4.4 years), the 146 isotope remains.   A typical shipment of promethium oxide contained about 3 ppm ^{146}Pm, which resulted in gamma activities in typical power sources of several millicuries.

MONITORING EQUIPMENT

A. Hand and Shoe Monitor

Most of the monitoring problems in handling ^{147}Pm are associated with the difficulty in detecting its soft beta emission. Since good contamination control requires the sensitivity of pulse counting techniques as opposed to continuous ionization current measurement, G. M. detectors are used in most portal monitors and hand and shoe counters, to meet this requirement economically. However, the metal or glass-walled G. M. tubes usually used in these devices have typical wall thicknesses of 30 to 40 mg/cm^2, much too thick for good efficiency at ^{147}Pm beta energies.

Since several manufacturers have recently made large-area thin-window G. M. probes available for portable survey meters, the author decided to use one type (the Eberline, Model HP-210, hand probe) in the construction of a shoe monitor. The detectors in these probes are of the "pancake" geometry, having thin mica windows (1.4 to 2.0 mg/cm^2) of relatively large area (15 cm^2). The ready-made probes also provide a rugged housing for the fragile detectors.

Two HP-210 probes, with handles removed and detectors connected in parallel, were mounted in the simple housing shown in Fig. 3. The probes were mounted on 7-inch centers under the foot rest to count the heel and mid-sole of the average shoe. Quarter-inch aluminum sheet was used in the sides and back of the housing, and 1/16 inch for the foot rest. The entire assembly was covered with 1/8-inch lead sheet to reduce background counts from low-energy gamma

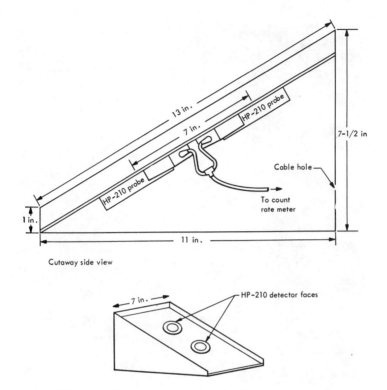

Fig. 3.   Promethium-147 Shoe Monitor.

emitters also handled in the area. It may be of interest to note that the manufacturers can also supply these probes with tungsten shielded cases. These may be preferred in areas where higher energy background is a problem.

The detectors in the shoe counter, and in a separate probe used for monitoring hands and clothing, all share a common 900-volt power supply and readout in a bench-type count rate meter (Fig. 4). With this arrangement, the background count rate was about 700 cpm in the work area.

Fig. 4. Hand and Shoe Monitoring Station. The
Large Instrument is the LRL Standard Alpha Hand
and Shoe Counter.

The G. M. detectors have a counting efficiency for ^{147}Pm betas of about 35% under "ideal" conditions. The calibration factor for the shoe monitor is approximately 2.5 x 10^3 cpm/nCi/cm^2 for ^{147}Pm uniformly distributed over the sole of a plastic shoe cover of the type used in this Laboratory.

B. Continuous Air Monitor

A small continuous air monitor (CAM), designed at the Lawrence Radiation Laboratory for radioiodine detection, was used for the promethium work.[1] The detector is a double-ended thin window G. M. tube with a window thickness of about 4 mg/cm^2. In operating the instrument, two filter papers were placed with the collecting sides facing the windows, and room air was drawn through them at about 1 cfm using the house vacuum system. An alarm trip circuit was set to alarm at predetermined levels above background. The detector counting rate was continuously monitored on a built-in meter and strip chart recorder. Since the detector head-filter holder assembly was mounted on the top of the instrument (Fig. 5), the detector was conveniently shielded against laboratory background radiation by stacking lead bricks around it.

The sensitivity of this instrument for ^{147}Pm air monitoring proved to be more than adequate. Calibration data indicated that an 8-hour exposure to ^{147}Pm at 0.1 x MPC$_a$ would produce a full-scale reading on the CAM.

Fig. 5.   Continuous Air Monitor for Beta Emitters.

C. Survey Meters

Contamination surveys were conducted with G. M. survey meters and end-window or "pancake" G. M. probes. A calibrated Nuclear Chicago Model 2588 "cutie pie" was used for high level monitoring of un-shielded sources.

D. Personnel Dosimetry

LiF powder thermoluminescent dosimeters, Harshaw TLD-700 (TLD) were used for finger dosi-metry and as a supplement to the LRL film badge for whole-body monitoring.

The assembly of the power sources involved delicate work which could not be handled remotely. During this phase of the project, the finger and body TLD's were read out daily for exposure control. The film badges were read out monthly.

OPERATIONAL EXPERIENCES

A. Cell Loading

To understand why exposure control was so criti-cal in this operation, one should review the details of a typical cell loading. All encapsulation operations were performed in a gloved box. The promethium oxide was loaded into aluminum capsules in 100-Ci batches. The promethium was preweighed and poured into a capsule, capped with a disc, soldered, and placed inside a larger aluminum capsule. Two layers of tape had been placed on the outer surface of the larger capsule as an aid to decontamination. After

loading, the capsule was sealed with solder, bagged
out of the box, and placed in a special portable enclosure that had been mounted in the hood. The first
layer of tape was then carefully removed. After de-
contaminating as much as possible, the capsule was
transferred out of the gloved box directly into the
hood. Decontamination continued with removal of the
second layer and further cleaning until the outside
surface of the capsule was free of removable activity.

The total loading and cleaning operation took about
2 hours. Since the 100-Ci batch read about 1000
rad/hr beta at the surface bare and 35-40 R/hr bremsstrahlung and gamma after being sealed in the second
disc, this operation accounted for 90% or more of the
total exposure received.

B. Spills

One small spill occurred while a wire was being
manipulated on the outside of one of the capsules.
Apparently a small amount of ^{147}Pm under the wire
was missed during decontamination. The material
fell on the floor in front of the hood and was stepped
on. It was first found when one of the experimenters
checked his hands and feet at the shoe monitor. A
few spots that ran about 1500 cpm above background
were found on the floor. The floor was easily decontaminated. Air sampling indicated no airborne activity.

A second incident occurred involving the temporary loss of some promethium, although it was not
technically a spill. A small 100-Ci source sealed in
a plastic bag had been placed in a can at the back of
the hood for temporary storage. The next morning
when work was to be resumed, the source could not

be found.   A thorough check of the room and trash
cans revealed nothing.   The source was finally found
in the hood filter upstairs in the fan loft.   Apparently
the hood draft had carried the source through the
ducting and deposited it on the front of the 1000-cfm
filter.   Once this possibility was considered, the
source was easily located by checking the duct and
filter with a portable survey meter.

C. Air Sampling

No activity above background was found on any of
the air samples taken during this operation.   As noted
above, the CAM had a detection capability of less
than 0.1 MPC hours.   Filter papers from the fixed
room air sampler which are counted once per day
verified the CAM results.

D. Personnel Exposures

Personnel rotation and daily readouts of finger
and body TLD's were successfully used in keeping all
personnel exposures below maximum permissible
quarterly levels.

SUMMARY

This project illustrated the need for flexibility
in designing a health physics program to fit the characteristics of the radionuclide and the requirements
of the experimenter.   Pre-planning and close coverage
by the health physics and monitoring staff have resulted in a successful and clean operation, with personnel exposures well within prescribed limits.

REFERENCES

1. S. Block, E. Beard, and O. Barlow, Health Physics 12, 1609 (1969).

2. C. M. Lederer, Table of Isotopes, 6th Edition, John Wiley & Sons (1967).

[181]TUNGSTEN CONTAMINATION INCIDENT*

D. Busick and G. Warren
Stanford Linear Accelerator Center
Stanford University, Stanford, California

INTRODUCTION

The Stanford two-mile accelerator has been in operation since May 26, 1966. During this time several unexpected health physics problems have been defined and workable solutions found. The major anticipated problems naturally are related to external radiation produced by the beam directly and the activation of beamline components. One problem, the extent of which was not fully realized before beginning operation, is radioactive contamination. The contamination problem has two origins: (1) machining activated beamline components and (2) cooling water. Of these two, cooling water has proven to be the most troublesome.

---

*Work supported by the U.S. Atomic Energy Commission.

The potential problems from irradiated water were examined during the design phase of SLAC by DeStaebler.[1] Based on DeStaebler's results, Coward calculated the saturation activity produced in a large water dump by a one (1) megawatt electron beam (Table 1). Of the four major isotopes produced, ^{7}Be

Table 1

Expected Saturation Activities Formed by
Stopping a 1 Megawatt Electron Beam
in Water

Daughter Nuclide	Half-Life	Saturation Activity (curies/MW)
O^{15}	2.1 min.	35,000
N^{13}	10.0 min.	1,390
C^{11}	20.5 min.	1,390
Be^7	53.0 days	280
H^3	12.3 years	400

and ^{3}H have the longest half-lives and should be the most troublesome to deal with, but ^{7}Be is very efficiently removed by mixed bed resins and the major problem is handling the contaminated resins with curie quantities of ^{7}Be. The production and removal of ^{7}Be has been examined experimentally by Busick.[2] Another source of contamination within the water system is the corrosion of irradiated elements that have become radioactive. Usually the contribution to the total activity is small when compared to the [7]Be present. One incident at the positron source demonstrated that corrosion can indeed be a serious problem.

DESCRIPTION OF INCIDENT

Positrons are produced by an electron beam striking a target called the positron source within the accelerator. The positrons are then heavily focused and accelerated through the remainder of the accelerator.

Typically an electron beam with an average power of 100 kW strikes the positron target. This power dissipation presents severe cooling problems. In addition, the solenoids and beam scraper require cooling. Figure 1 shows a schematic of the water flow and equipment layout at the positron source.

The Health Physics Group routinely samples the cooling water and sump water for this area and [181]W had been detected in small amounts. These samples contained concentrations of 1×10^{-4} μCi/cc down to 3×10^{-6} μCi/cc. The edge cooled coils inside solenoid A had shorted and were scheduled to be replaced by center cooled coils during a two-week maintenance shut-down period. This meant that the water lines had to be broken and activated parts removed.

Like many unexpected events of this nature, we discovered the problem quite by accident. On October 30, 1967, a fire was reported at the positron source area. This occurred at about 1100 hours. Also that same morning an employee was engaged in cleaning and reworking the positron source sump. Members

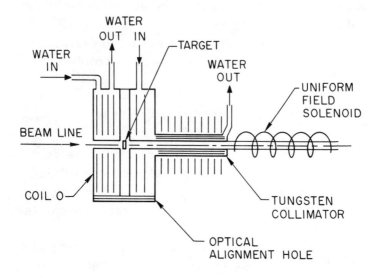

Fig. 1. Positron Source.

of the Health Physics Staff investigated the fire and
made the necessary radiation measurements, then pro-
ceeded to the sump from which an employee was observed bailing water. We found that his shoes were
contaminated as well as the sump lid and aisle.

At this point it became obvious that the source
of the contamination was the e+ source. A maintenance crew had started disassembling the e+ source
that morning. Radiation surveys were made and the
health physics emphasis had been placed on external
radiation monitoring. This was true because analysis
of water samples from this water system were low
($3 \times 10^{-6} \mu Ci/cc$) and the γ radiation levels from the
activated e+ source, a copper target, were~5R/hr.
It was our thought at the time that as long as the tungsten stayed with the water there would be minimal contamination resulting from working on the hardware.

The source of contamination was apparently insoluble deposits found inside of the solenoid, including the coils and associated outlet cooling lines. The deposits appeared as a fine, reddish colored film. The contamination began to appear shortly after a low point in the solenoid was drained or following removal of the cooling lines from the solenoid. The insoluble material inside of the solenoid was dislodged, contaminating the physical structures in the immediate vicinity of the solenoid and the people doing the work. Micro-curie quantities of ^{181}W were found on shoes, clothing and hands.

The logical source of the ^{181}W and ^{185}W is believed to be the tungsten collimator just down beam of the e^+ target. A gamma scan of resin samples from the water system revealed the presence of ^{181}W, ^{7}Be, ^{57}Co, and ^{58}Co. The cobalt isotopes are thought to be from the protective nickel plating of the collimator in the system. The nickel plating is apparently descrepitating, allowing oxidized tungsten to be released to the water cooling system.

SURVEY TECHNIQUES

From a health physics viewpoint, this incident points out the inherent risk associated with dependence on water sampling data and the difficulty in using portable survey instruments to assess contamination potential in this situation. The principal isotopes involved were ^{181}W, ^{185}W, and ^{7}Be. The detection of ^{7}Be and ^{181}W is difficult at the arbitrarily set but acceptable contamination level of 1×10^{-4} Ci/cm^2.

Table 2 illustrates the response of three survey meters to distributed solution sources of ^{181}W and ^{7}Be. ^{185}W can easily be detected with the GM portable

Table 2

Relative Response Above Background of Selected Radiation Monitors to 60 keV and 477 keV Gamma Energies

Survey Meter		^{181}W	^{7}Be
G. M.	mR/hr - μCi	0.2	0.02
Ion Chamber	mR/hr - μCi	0.03	0.01
CRM - G. M.	c/m - μCi	3×10^3	8×10^2

survey meter since it emits a beta particle. ^{181}W emits a 60 keV x-ray while ^{7}Be emits a 477 keV gamma ray 12% of the time.

The SLAC Health Physics Group has a variety of laboratory instruments including capability for pulse height analysis. This instrumentation is necessary for our operational and research functions. By using this laboratory equipment we were able to assess the extent of the contamination problem quickly.

The people involved in this incident were contaminated externally but were not exposed to detectable internal contamination. The maximum permissible body burden for the individual isotopes involved are higher than the amounts detected on their body surfaces and clothing. Personnel air samplers did not reveal the presence of airborn radioactivity attributable to this incident.

CONCLUDING REMARKS

The fact that a radioactive contamination problem occurred at an AEC installation is probably not surprising. However, high energy accelerators do not represent typical AEC installations; SLAC does not possess large inventories of fission products or activated materials.

Due to the operating characteristics of this machine, such as low activation cross sections (microbarns), and limited types of solid target material, one would not expect residual radioactivity in the form of removable contamination to be a major problem. In fact, it resolves usually into more of a nuisance problem in controlling the many relatively low-level sources of radioactive materials (tools, wiring, plumbing, etc.).

What makes this particular problem both interesting and difficult to assess are the kinds of radioactivity produced. Normal operational health physics methods are not adequate at the usual levels of detection. Some of the radioelements produced by high energy accelerators are essentially pure gamma emitters and will not be detected at the arbitrarily set but acceptable levels of radiocontamination prevalent in this country. In the absence of corpuscular emission, ion chambers and Geiger counters are not sufficiently sensitive to detect existing contamination limits. Portable scintillation counters are not always effective because of their high sensitivity to existing low energy background radiation.

To the aforementioned problem we must add the high gamma radiation fields, from induced radioactivity in the targets and associated accelerator hardware, which ranges from a few mR/hr to R/hr levels at reasonable working distances.

The difficulties mentioned above leave us with a choice of laboratory-type instrumentation for detecting radiocontamination. Ideally this instrumentation would be a multi-channel analyzer and NaI crystal for rapid identification and quantity determination.

It is doubtful if we would have fully realized the magnitude of this incident in time to minimize the spread of radioactivity without the laboratory equipment mentioned. This is true in part because radiocontamination was not thought to be a problem at SLAC.

Ten men were directly involved in this incident. One man was heavily contaminated (~50 μCi). The remainder had < 5μCi of γ-activity on their clothing and hands which could have been easily overlooked with conventional health physics measurements. The men were successfully decontaminated and some clothing was confiscated.

All personnel involved in this incident were subsequently counted in a mobile whole-body counter and found to be free of those radionuclides attributable to SLAC operations.

REFERENCES

1. H. DeStaebler, "Photon Induced Residual Acticity," Report No. SLAC-TN-63-92, Stanford Linear Accelerator Center, Stanford University, Stanford, California (1963).

2. D. Busick, "^{7}Be Build-Up in a Large Water Beam Dump System at the Stanford Linear Accelerator Center," Report No. SLAC-PUB-521, Stanford Linear Acclerator Center, Stanford University, Stanford, California (1968).

USE OF LOCAL SHIELDING IN
EXPERIMENTAL CAVES AT A CYCLOTRON

William W. Wadman III
University of California
Irvine, California

INTRODUCTION

At Lawrence Radiation Laboratory, Berkeley, we were confronted with the following problem:

Shield a very large experimental set-up at a medium energy cyclotron to afford the optimum hazard protection with a minimum of shielding and surveillance instrumentation.

We attacked the problem at several sides. 1) Did a literature search to utilize information already in existance. 2) Determine the approximate hazard based upon the parameters specified by the chief researcher. 3) Perform research in the areas where insufficient data existed from which we could design the shielding. 4) Utilize off the shelf electronic instrumentation where radiation safety and surveillance equipment was required. 5) Perform follow-up radiation surveys and recommend shielding changes as required, after first operational tests were made.

METHOD OF ATTACK—NEUTRON YIELD DETERMINATIONS

Lets briefly discuss what each step of the attack entailed, then return to the preliminary design configuration for the large cave. The literature search revealed data which went into the family of curves shown in Fig. 1.[1-13] Note that the total neutron yield increases approximately as a cubic function of the energy. Note the difference in yields between types of ions, and target materials for the same ions. Table 1 gives the ion-energy ranges and the total neutron yield at 5 ma at the maximum energy listed for that ion. These are values calculated from the first curves.

Out of this data we were able to gain considerable information. However we were still lacking the neutron spectrum or the neutron relaxation lengths or the angular dependence of the neutron yields.

ATTENUATION AND ANGULAR YIELD STUDIES

We employed neutron reaction threshold detectors to study the fast neutrons—the secondary radiation produced by stopping the beam ions—and the primary radiation hazard here. The properties of importance to us are shown in Table 2.

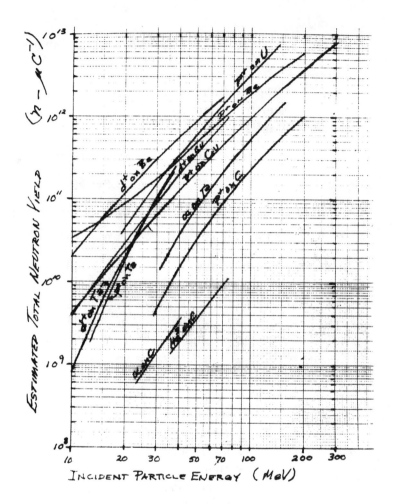

Figure 1.

Table 1

88-Inch Cyclotron Beam Energies and Maximum
Neutron Yield

Particle (Ion)	Lowest Energy (MeV)	Highest Energy (MeV)	Maximum Yield (4π) n/sec-5 ma
$_1H^{1+}(p^+)$	5.0	50	2.5×10^{15}
$_2H^{1+}(d^+)$	10.2	60	3.0×10^{15}
$_3H^{1+}(t^+)$	15.5	150	(Not to be accelerated)
$_3He2$	15.5	150	3.0×10^{15}
$_4He2(\;^{++})$	21.5	120	3.7×10^{15}

From each set of reactions we were able to compare reaction activies as a function of depth in concrete the five energy groups and satisfy ourselves in part to the validity of the data from the parallel attenuation curves of each, Secondly, from the activities of the reaction products, we were able to calculate the approximate neutron spectra.[14,15]

Figure 2 shows the threshold detector reaction cross sections as function of neutron energy.[15]

By placing detectors radially around the target, we could determine the angular yield of neutrons by energy groups as shown in Fig. 3.[16]

We placed the detectors in precision cast concrete shielding in order to study the neutron attenuation. We were thus able to determine the effectiveness of concrete for neutron shielding and gain some detailed information of the changes, if any, in the neutron spectrum with depth of penetration in the shield.

Table 2
Threshold Detector Properties

Reaction	Calculated Threshold (MeV)[a]	Peak Cross Section (Barns)	Product Half-Life	Target Isotope (%)	γ-Ray Energy of Product (MeV)
$^{58}Ni(n,p)^{58}Co$	1.1	0.556	71 days	67.8	0.81
$^{27}Al(n,a)^{24}Na$	6.7	0.243	15 hours	100.0	1.37, 2.74
$^{203}Tl(n,2n)^{202}Tl$	8.5	2.78	12 days	29.5	0.44
$^{58}Ti(n,2n)^{57}Ni$	12.4	0.25	37 hours	67.8	1.36
$^{203}Tl(n,4n)^{200}Tl$	24.7	$\simeq 1.3$	27 hours	29.5	0.368, 1.207

[a]The energy at which the cross-section value is 2% of the maximum cross-section value.

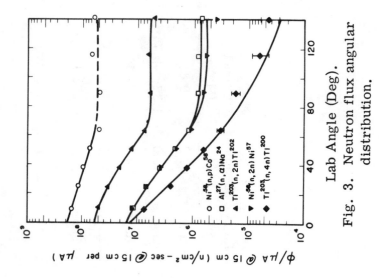

Fig. 3.   Neutron flux angular distribution.

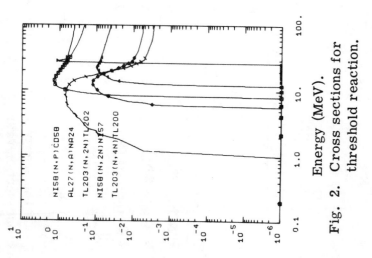

Fig. 2.   Cross sections for threshold reaction.

We also studied the effects of replacing the first foot of concrete with iron and depleted uranium for the same exposure geometry. An example of this is given in Fig. 4. Those details have been reported elsewhere. Figure 5 shows a set of attenuation curves for the iron-faced concrete.[17] Note that the spectral and attenuation curves are inverse square connected.

From the experiment, we were able to determine the neutron yield ratios for the three ion-beams of the maximum energy proposed for use in the new cave. We were able to determine the neutron relaxation lengths for the neutrons from the three ion beams and the effects of the introduction of heavy interface materials. And we were able to measure angular dependence of the neutron yield, to some degree, for each ion.

PRELIMINARY DESIGN CONSIDERATIONS

Figure 6 is a plan view of the 88" cyclotron and its associated experimental caves. The caves of interest to us now are caves 3 & 4. We were asked to specify shielding to permit beam losses of 10 μA in Cave 3, 1 μA just past the 110° magnet and in Cave 4, and about 0.1 μA past the 2nd 110° magnet and 1.0 μA in the Faraday Cup. Armed with the experimental data and the assurances of the experimentor that these parameters of beam loss were "absolute maxima" and very unlikely to be achieved, the shielding was designed and barricades set.

A remote neutron sensor with associated electronics to activate upstream beamstops was assembled. The system is a modified version of the fast shut off

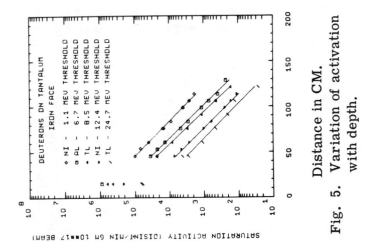

Fig. 5. Variation of activation with depth.

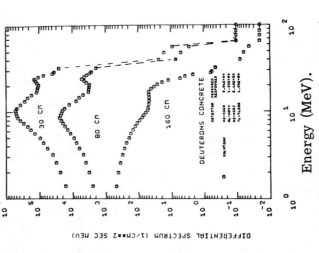

Fig. 4. Energy spectrum variation with depth in a concrete shield.

system reported in UCRL 17728, presently in use a the 88" Cyclotron.[18]

When the total shielding compliment was in place as shown in Fig. 7. and the cave operated at the parameters specified in the original problem, the maximum radiation field measured in an unbarricaded area was 7.5mrem/hr.

However, beam transport was improved, and we were faced with problems. Let me say that the problems created by a 10 fold increase in beam loss at the two high-loss areas were not insuperable. An extension of the added thickness of roof shielding over cave #3 and specification of an iron "dog house" over the analyzing slit would accommodate the increase.

Along with that recommendation we also suggested replacing the heavy tantalum slits with carbon to reduce the neutron yields. Carbon facing of most of the face area of the slits was done, proportionally reducing the neutron yield. This project is still under study, and I might add, still experiencing tolerable levels of radiation exposure.

SUMMARY

In conclusion, local shielding is practical to use at particle accelerators, if done prudently. Permissible radiation levels may be achieved by specifying maximum beams currents to any high-loss area.

Additional shielding requirements must be met when the need is demonstrated. Accelerator parameter changes require a new radiation survey and possibly shielding changes.

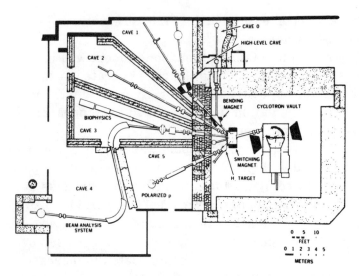

6. The 88" cyclotron and
    experimental caves.

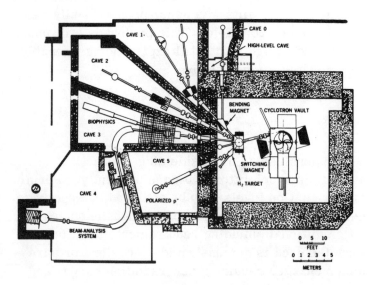

7. The cyclotron fully
    shielded.

Two drawbacks to locally shielding such a high intensity situation. 1) It requires much greater continuous attention by a health physicist   2) Support personnel find it psychologically unsettling to be able to view a beam pipe known to contain beam ions.

ACKNOWLEDGEMENTS

My thanks to Mr. James DeGrafenread for supporting my paper although it is not directly connected with my present duties, and to Mrs. Margie Alexander for taking time from her ordinary duties to type this paper.

REFERENCES

1. A. J. Allen, J. F. Nechaj, K. H. Sun, B. Jennings, Physical Review Vol. 84 #4 Feb. 1951 (15MeV d^+ & 30MeV Alpha)

2. F. Ajzenberg-Selove, C. F. Osgood, C. P. Baker, Physical Review Vol. 116 #6 Dec. 1959 (10.5 MeV p^+).

3. W. E. Crandall, G. P. Millburn, L. Schecter, Journal of Applied Physics Vol. 28 #2 Feb. 1957 (12 MeV P^+ & 24 MeV deuterons).

4. L. W. Smith, P. G. Kruger, Physical Review Vol. 83 #6 Sept. 1951 (10 MeV d^+).

5. E. L. Hubbard, R. M. Main, R. V. Pyle, Physical Review Vol. 118 #2 April 1960 (10.4MeV/nucleon ^{12}C, ^{14}N, ^{20}Ne).

6. J. H. Gibbons, R. L. Macklin, Physical Review Vol. 144 #2 April 1959 (5MeV P$^+$ & 9MeV Alphas.)

7. J. M. Cassels, T. C. Randle, T. G. Pickavance, A. E. Taylor, Correspondence XXI. Serial 7, Vol. 42 #325 (1951).

8. Y. K. Tai, G. P. Millburn, S. N. Kaplan, B. J. Moyer, Physical Review Vol. 109 #6 March 1958 (18 & 32 MeV p$^+$).

9. W. W. Wadman, Health Physics Vol. 11, #7 Pg. 659 (40 & 80 MeV alpha).

10. R. R. Borchers, R. M. Wood, Nuclear Instruments & Methods Vol. 35 (1965) (8-14 MeV [1MeV steps] d$^+$.

11. T. W. Bonner, A. A. Kraus, Jr., J. B. Marion, J. P. Schiffer, Physical Review Vol 102 #5 June 1956 (1.8 to 5.3 MeV alpha).

12. E. Tochlin, G. D. Kohler, Health Physics Vol. 1 #3 Dec. 1958.

13. W. J. Knox, UCRL 1398 July 1951 (190MeV d$^+$ 340 MeV P$^+$).

14. A. D. Kohler, UCRL 11760 1964.

15. J. Routti, CDC 6600 Program SAMPO 1966.

16. W. W. Wadman III, Nuclear Science and Engineering, Vol. 35 #2 Pgs. 220-226 (1969).

17. W. W. Wadman III, A. J. Miller, A. R. Smith, J. Routti, UCRL 18050 1968 and AND/CNA Transactions, Vol. 11, #1, June 1968.

18. H. S. Dakin, W. W. Wadman, UCRL 17728 May 1967.